T0186517

Advances in
Cognitive Science

AAAS Selected Symposia Series

Advances in Cognitive Science

Steps Toward Convergence

*Edited by Manfred Kochen
and Harold M. Hastings*

Routledge
Taylor & Francis Group

LONDON AND NEW YORK

AAAS Selected Symposia Series

First published 1988 by Westview Press

Published 2018 by Routledge
52 Vanderbilt Avenue, New York, NY 10017
2 Park Square, Milton Park, Abingdon, Oxon OX14 4RN

Routledge is an imprint of the Taylor & Francis Group, an informa business

Library of Congress Catalog Card Number: 87-34624
ISBN 978-0-813-37471-0

ISBN 13: 978-0-367-01425-4 (hbk)

About the Series

The *AAAS Selected Symposia Series* was begun in 1977 to provide a means for more permanently recording and more widely disseminating some of the valuable material which is discussed at the AAAS Annual National Meetings. The volumes in this *Series* are based on symposia held at the Meetings which address topics of current and continuing significance, both within and among the sciences, and in the areas in which science and technology have an impact on public policy. The *Series* format is designed to provide for rapid dissemination of information, so the papers are reproduced directly from camera-ready copy. The papers are organized and edited by the symposium arrangers who then become editors of the various volumes. Most papers published in the *Series* are original contributions which have not been previously published, although in some cases additional papers from other sources have been added by an editor to provide a more comprehensive view of a particular topic. Symposia may be reports of new research or reviews of established work, particularly work of an interdisciplinary nature, since the AAAS Annual Meetings typically embrace the full range of the sciences and their societal implications.

ARTHUR HERSCHMAN
Head, Meetings and Publications
American Association for the
Advancement of Science

Contents

Tables and Figures

Acknowledgments

Dr. Yates (Chapter 3) thanks Donald Walter and Kirstie Bellman for many helpful discussions and ideas. He has tried to capture some of their contributions, in paraphrase, but takes full responsibility for the opinions expressed in his chapter.

Drs. Reeke and Edelman (Chapter 4) thank the Henry J. Kaiser Family Foundation and International Business Machines Corporation for their support of their research.

Drs. Hastings and Waner (Chapter 5) thank Drs. M. Conrad and R.M. May for helpful discussions, and Gretchen Hastings, Bruce Higer, Alex Hoyos, Susan Kenney, Janet Kerner, Steven Rosenthal, Dr. John Weidner, and Carmine Zingarino for programming.

Dr. Greenspan (Chapter 8) thanks Drs. John Bonner and David J. Grunwald for helpful discussions. The author's research cited in the chapter was supported by grants from the U.S. Public Health Service, the Searle Scholars Program, the McKnight Foundation, and the Alfred P. Sloan Foundation.

Dr. Freeman's (Chapter 9) research was supported by grants MH 06686 from the National Institute of Mental Health and NS 16559 from the National Institutes of Health, U.S. Public Health Service.

Manfred Kochen
Harold M. Hastings

1. Introduction

The collection of papers brought together in this volume is the result of two symposia held at the annual meeting of the American Association for the Advancement of Science (AAAS) in May 1985 in Los Angeles. The first symposium reported recent advances in our understanding of the relation between brain and brainlike structures and mind or mindlike functions to a broad audience of scientists and others interested in cognition, information, neuroscience, and artificial intelligence. The second symposium shared explorations in evolutionary computing with scientists, engineers, and others interested in machine intelligence or in modeling biological intelligence with the help of computer metaphors.

In his lecture opening the AAAS meeting, Murray Gell-Mann, Professor of Physics at the California Institute of Technology, singled out cognitive science to illustrate the convergence of mathematics, computer science, biology, and engineering. He noted striking similarities between learning and evolution: the two fields share a central concern with the emergence of surface complexity from a deep, underlying simplicity, as well as the emergence of self-organization. Gell-Mann stressed the importance of random inputs to automata that are made reliable by attractors, as these are understood in mathematical synergetics (Haken 1983).

In another lecture at the meeting, Daniel Koshland, Jr., Professor of Biochemistry at the University of California at Berkeley and editor of the AAAS journal, *Science,* pointed out striking similarities between neuronal and molecular mechanisms that play a role in the emergence of memory, in adaptation, and in stimulus integration. The human immune system has also been observed to exhibit memory. Imaginative designers of newer kinds of machines expect them to achieve goals by variation and selection, to learn to use resources efficiently and from correctable mistakes. Adaptive systems have been modeled (Kochen

1971) that select their own goals in a state-space to which a function assigning values has been given.

The concepts of computer science are no longer used solely as metaphors for thinking about the nature of information and learning. The existence of hard data, of replicable rigorous methods of inquiry, and of more advanced theoretical underpinnings that are beginning to yield insightful results offers hope for the emergence of clean, real science in this area.

The study of cognition has a long history. It is concerned with such questions as:

- How can a learner observe its own actions?
- How can it classify objects and events of its experience into conceptual categories?
- How can it recall and record such coded experiences?
- How can it anticipate previously unencountered experiences in an integrated way?
- How can it control shifts in attention?
- How can it pose questions?
- How can it decide and plan?
- How can it represent knowledge so that it can cope with an increasing variety and number of problems?

This volume should enable the reader to assess how close cognitive science is to an idea of clean science, whether enough is known, and whether we are mature enough to organize research and penetrate in a fruitful way to the essential issues.

Chapter 2 deals with evolutionary biology. The concepts of species, their adaptiveness, Darwinian fitness, and natural selection have influenced thinking about cognition since the concept of the computer first stimulated the imagination of scientists. Indeed, even before molecular biology achieved its great successes by explicating with great precision the mechanisms of genetics and their role in evolutionary theory, Baricelli (1957) had simulated the evolution of numerical organisms. He simulated all the genetic mechanisms known in the 1950s (e.g., mutation, crossover) on the computer at the Institute for Advanced Study in Princeton, N.J. (a 1K machine using the Williams electrostatic storage tube!) to demonstrate the emergence of new stable patterns after many nights of computer runs. Hailed informally by von Neumann and Courant as important pioneering work, Baricelli's work was the first of many similar subsequent simulations that have never properly credited his initial effort, and whose authors were probably unaware of it.

The issue raised in Chapter 2 is whether nature can be regarded as making decisions. For example, how was it decided whether certain species of birds should nest in trees or on the ground? If such decisions were analyzed mathematically, as if nature tried to optimize fitness, the rules used have a striking resemblance to those of classical decision theory in sufficiently simple models. In Chapter 2, William Cooper shows us that: (1) the simplest population process model implies the axioms of von Neumann and Morgenstern; and (2) a more complex population process model implies the axioms of Savage for a theory of decisionmaking based on subjective probabilities and utilities. Can a deeper understanding of biology instruct us about how to reason? The answer depends on how we interpret and extend the fitness function. The fitness function refers exclusively to the multiplicative factor by which a population grows during an ordinary breeding season, which, in turn, depends on the number of females reaching reproductive age, having offspring, the survival of those progeny, etc.

By describing how nature causes certain choices to be adopted by supposing that it uses decision rules analogous to those employed by human decisionmakers, we sometimes imagine that nature, or the biological systems it causes to evolve, calculates or computes. Psychologists often make the implicit and unexamined assumption that the brain somehow computes with its inputs. In Chapter 3, F. Eugene Yates raises the question of whether biological systems compute.

In Chapter 4, George N. Reeke, Jr. and Gerald M. Edelman continue the theme of Darwinian evolution by asking whether a principle analogous to Darwin's could operate in somatic time. Their selectionistic approach does not assume pre-existing categories in the environment. The organisms they study are not assumed to know what problems to solve; their only goal is to survive in a highly uncertain environment. To make this possible, nature takes advantage of variability. Neuronal circuits are selected during an embryogenetic first stage. At a later stage, there is selection among diverse, pre-existing systems: the "wiring" of neuronal groups, which are the functional units in the nervous system, is not preprogrammed, but can adapt to the environment by means of differential amplification of selective connectivity.

Information cannot be transmitted faster in a neural net than the neuronal speeds specified by Helmholtz (Boring). This would make it virtually impossible for the nervous system to perform double-precision or matrix arithmetic. The nervous system is a network, with many functionally specialized regions and extensive overlap in arborization, and with no single neuron indispensable. Nor is the exact connectivity genetically specified. Neuroscience has thus provided computer science with the concept of concurrent or parallel computation and connectionist models.

Chapter 5, by Harold Hastings and Stefan Waner, continues the theme of evolutionary adaptivity, connecting it with ideas derived from thermodynamics and statistical mechanics that led to the proposal for "Boltzmann machines." It shows that analogues of annealing—in which a small degree of randomness in a system prevents it from falling into the traps of local minima as its potential energy and random noise decrease with time—and "soft programming" are sufficient for evolutionary learning. "Soft programming" refers to a modification of the system's potential function through feedback from and interaction with the environment. Three principles are proposed as governing the operation of evolutionary learning systems:

1. System trajectories pass arbitrarily close to all states of the system.
2. Satisfactory equilibria can be attained if the temperature is decreased slowly enough.
3. System dynamics are based on soft programming.

Chapter 6 explores more fully, in the context of computer science, the pervasive biological phenomenon of massive parallelism. Noting that molecular assemblies that function as higher order biological systems are good at reasoning and recognition, while serial electronic systems are good at arithmetic and sorting, H.L. Resnikoff shows that connectionist theories of memory and learning synthesize ideas from information theory and concurrent computation to proposed machines that could perform like biological and electronic systems. A parallel computer has been developed that may serve cognitive scientists such as Uhr (1983) as a useful tool.

Chapter 7, also by Waner and Hastings, describes a neural net model that realizes evolutionary learning. Again, processes observed in natural evolution are transplanted to computer science.

In Chapter 8, Ralph J. Greenspan returns us to biology. He inquires into the nature of genetic variation in brain ontogeny over species, as well as of the species-specific structure of the nervous systems in individuals. The great deal of individual identity in each neuron relative to other tissues and organ systems makes possible higher order mental processes. Using data from developmental mutations in nematodes, fruit flies, zebrafish, and the house mouse, Greenspan finds four major classes of genes that play a role in the development of nervous systems. As we move up the phylogenetic scale, the increases in brain complexity and corresponding functional levels seem to be due to greater genetic efficiency or versatility in brain development.

Chapter 9, by Walter Freeman, delves into a particular and central issue of neurobiology: the biological basis of mental images. He shows that rabbits, confronted with olfactory stimuli, react to their expectation of an odor rather than solely to the stimulus. It is a remarkable synthesis of advanced neurophysiology, the mathematics of synergetics, and the sophisticated use of computer technology for both experimentation and modeling.

Having progressed from concepts in evolutionary biology in Chapters 1 and 2 to the new ideas in computer science (Chapters 3–7) that they inspired, and then returned to biology in Chapters 8 and 9, we now turn once again to new ideas in computer and information science.

Chapter 10 by Michael Conrad analyzes evolutionary learning through a novel approach called evolutionary programming. Evolutionary programming contrasts sharply with conventional computer programming in that (1) emergent primitives play a key role in evolutionary programming, (2) evolutionary programming is adaptive, and (3) evolutionary programming makes efficient use of computational resources. Chapter 11, by Conrad, Roberto R. Kampfner, and Kevin G. Kirby, describes simulation experiments using a reaction-diffusion approach to evolutionary learning. This approach shows how a "conventional" information-processing system can be structured for evolution. Chapter 12, authored by Kampfner, models intra-neuronal dynamics as a reaction-diffusion system and shows how this modeling explains evolutionary learning.

The remaining chapters deal with aspects of this issue. In Chapter 13, Ray Solomonoff—who founded algorithmic complexity theory independently of Kolmogorov and Chaitin—sheds important new light on the problem of inductive inference, which he has studied for the past three decades. He shows how algorithmic complexity theory, in combination with computational complexity theory, can be brought to bear on the design of evolutionary systems that learn problem solving from experience. At the heart of his approach is the probabilistic representation of knowledge about solved problems, an important area for future research.

Chapter 14, by Robert Lindsay and Manfred Kochen, deals with another issue of problem solving: how to control backtracking in searching for a solution, and how to recognize novelty. Therein lies the essence of discovery, and the genesis of new ideas.

Kochen's Chapter 15 presents a new approach to the representation of knowledge that incorporates many of the features discussed in the previous 14 chapters, while the final chapter by Hastings and Kochen pulls together the hypotheses in computer science and evolutionary biology to argue that fruitful syntheses between the two are now likely, and that such a fusion of these two disciplines is leading to important advances in the cognitive sciences.

Bibliography

Baricelli, N.A., 1957: Symbiogenetic evolution processes realized by artificial methods, *Revista Methodos* 9:35–36.

Haken, H., 1983: *Advanced Synergetics*, Springer-Verlag, Berlin.

Kochen, M., 1971: Cognitive learning processes—an explication, in N.V. Findler and B. Metzler (eds.), *Artificial Intelligence and Heuristic Programming*, Edinburgh University Press, Edinburgh, Scotland.

Uhr, L., 1983: *Algorithm-structured Computer Arrays and Networks: Architectures and Processes for Images, Percepts, Models, Information*, Academic Press, New York.

2. Is Decision Theory a Branch of Biology?

Decision-theoretic Background

The classical theories of rational choice under risk and uncertainty, as exemplified by the systems of Ramsey (1926), von Neumann and Morgenstern (1953), Savage (1972), and Jeffrey (1983), give the impression of being largely self-contained. They are usually presented not as logical extensions of other theories, but instead as resting on independent conceptual foundations of their own. Modern decision theory is not usually classified as a branch or corollary of something else, in other words. It is true that certain other theories—in particular, certain important theories of subjective probability—are thought to stem from or form a portion of decision theory (Kyburg and Smokler 1980). But decision theory as a whole—the unified theory of how probabilities and utilities combine to control choices—seems to be widely perceived as something that cannot properly be founded upon any other known science.

This perception could be described in methodological terms as a presumption that decision theory is, so far as anyone presently knows, *irreducible*. Roughly speaking, one theory is by definition "reducible" to another if and only if its primitive concepts are definable in terms of the concepts of the other and its fundamental postulates are logically derivable from the postulates or theorems of the other (Nagel 1961; Schaffner 1977). Thus, for example, much of chemistry is thought to be reducible to physics, and much of classical genetics to the chemistry of DNA. The technicalities of reducibility need not concern us here, but it may be well to note that the question of the reducibility of a subject matter is independent of the issue of its historic origins. Thus, certain modern theories of decision were originally motivated by problems arising in economics, but it does not follow from this that they are reducible to general, independently established economic laws.

Some 28 different axiomatizations of the normative theory of decision under uncertainty are now available (for survey, see Fishburn 1981). These various theories differ in important details, but coincide in their general intent to formalize the process of rational choice making in terms of the maximization of subjectively expected utility. All are defended as being either intuitively rational or psychologically plausible. None are defended on the grounds that they are logical consequences of more fundamental scientific laws. Why this neglect of the possibility of an external scientific basis for decision theory? One likely reason is that when the emphasis is on rationality, attention turns naturally to such logical issues as internal consistency and coherence, and this emphasis makes decision theory appear to be, like deductive logic or mathematics, an abstract structure of pure reason whose validity is quite independent of that of any empirical science. And logical truths are, as everyone supposedly knows, to be sought exclusively in the faculty of rational intuition, not from external empirical disciplines.

Even in the case of non-normative, or descriptive, decision-theoretic studies, there has been little attempt to build up systematically from the known laws of other sciences. Psychology might appear to be an exception, and indeed, the experimental techniques of behavioral psychology have been applied extensively and successfully in the analysis of actual choice behavior (for surveys, see Edwards 1954; Lee 1971; Slovic, Fischhoff, and Lichtenstein 1977). However, it could be questioned whether very many of the descriptive hypotheses that have emerged from this activity were actually derived directly from more general psychological laws established previously. Certainly the basic notion of subjective expected utility maximization, the starting point for many descriptive theories, was not a prior discovery of experimental psychology.

Biological Background

Another science—evolutionary biology—has also entered the decision-theoretic scene to the extent that the biological adaptiveness of particular patterns of choice behavior is sometimes discussed (see Einhorn and Hogarth 1981 for review). But here again, it is hard to find anything in the decision-theoretic literature that could pass for a direct and explicit derivation of abstract decision rules from particular evolutionary models, or at least anything where the resulting rules are on the order of generality of the principle of expected utility maximization. And just as it apparently has not occurred to most decision theorists to attempt to find a direct connection between their foundational concerns and the details of evolutionary biology, so most contemporary biologists have shown little interest in the foundations of decision theory.

True, biologists have often *used* decision theory. Indeed, along with game theory, decision theory has recently come to play an increasingly prominent role in evolutionary analysis (beginning with Lewontin 1961; see Maynard-Smith 1978 for survey). However, the prevalent attitude seems to have been that the techniques of decision theory are applied *to* the biological problem, whose analysis is therefore an application *of* decision theory. The alternative possibility that the entire theory of decision might itself be inherent in the biological subject matter seems never to have been discussed seriously. The upshot is an apparent attitude among most biologists that there is nothing extraordinary about the use of decision theory in evolutionary biology, that applying decision theory in biology is much like applying calculus in physics or statistics in psychology. The usual paradigm of a logical calculus independent of, but applied to, an empirical problem is taken for granted.

But is it safe to assume that the biological application of decision theory is merely scientific business as usual? Something about the character of the evolutionary uses of decision theory suggests that it is not. For upon studying typical evolutionary analyses involving decision theory, one gains the impression that decision theory is not an imported analytic technique at all, but was already inherent in the biological problem structure. Its so-called "application" consists in a *renaming* of pre-established biological concepts using the vocabulary of decision theory. Thus, "environments" are rechristened "states of nature," "environmental frequencies" are called "prior distributions," Darwinian "fitness" becomes decision-theoretic "utility" (or the negated loss-function, etc.) and so forth (Templeton and Rothman 1974, 1978). More striking still is the fact that even the mathematical calculations required for the decision analysis are often identical with the standard evolutionary calculations that would have been used anyway had decision theory not been employed. In such cases, the *only* essentially new contribution made by introducing decision theory into the analysis is the reinterpretation in decision-theoretic terms of ordinary evolutionary reasoning and computation.

Especially curious is the fact that decision theory is centrally concerned with an agent called the "decisionmaker" and the behavioral choices he will or should make, while the evolutionary analyses that involve decision theory are about an entity called the "organism" (or "organism-type," "genotype," etc.) and the behavioral or other properties the organism can be deduced to possess. In a biological analysis based on decision theory, these two agencies are equated (i.e., under the decision-theoretic reinterpretation, the organism *becomes* the decisionmaker). Ordinarily, there is no analogy between a logical or mathematical formalism and the subject matter to which it is applied; for instance, the Newtonian differential calculus can be applied to planetary orbits, but is not itself explicitly about anything similar to either planets or orbits. Can it be pure coincidence that the formalism of decision theory is explicitly focused upon the same sort of entity that is also the object of the evolutionist's concern?

Such considerations suggest that decision theory may in some sense be directly reducible to evolutionary biology, and that, moreover, much of this reduction has already been unwittingly accomplished by the independent efforts of decision theorists and biologists.

An Illustrative Evolutionary Analysis

To judge intelligently whether a genuine reduction of decision theory to biology might be possible, it is necessary to examine the details of a typical piece of decision theoretically interpretable evolutionary reasoning. For the sake of a simple example, let us imagine a hypothetical organism which is capable of building its nest either on the ground or in a tree. The advantage of nesting on the ground is that food and living space are abundant there, guaranteeing that the organism will be able to raise two offspring to maturity if the nest is undisturbed. The disadvantage of a ground nest is that it is vulnerable to predators; if a predator finds the nest, neither offspring will survive. For definiteness, assume the probability of such predation is 3/8. Nesting in a tree, which is cramped but provides absolute safety from predators, assures that one and only one offspring will be raised to maturity. *Problem*: Which behavior does evolutionary theory predict would be found in nature, the ground-nesting or the tree-nesting?

To avoid biological complications, it will help to introduce some strong simplifying assumptions. Let us suppose that reproduction is asexual (haploid), and that the behavioral character of interest is controlled at a single locus of the genome at which either a gene for ground nesting or a gene for tree nesting may be present. Except for mutations, the offspring of members of the ground-nesting genotype will then be ground nesters, and the offspring of tree nesters, tree nesters. Let us suppose also that the population is very large, that each individual can breed only once in its lifetime (semelparity), and that the breeding is seasonal so that all individuals produce their offspring in lock step at the same age and time of year. The predation probability of 3/8 is assumed to be constant and independent from individual to individual and from season to season. Some neutral regulatory mechanism may be postulated which prevents the population size from growing indefinitely large; for instance, it may be supposed that when a certain population size ceiling is reached, both the ground-nesting and tree-nesting subpopulations are halved by disease or some other impartial overpopulation-induced disaster. Finally, let us assume that it is appropriate to adopt as a measure of genotypic fitness the commonly used textbook measure—the multiplicative population growth rate (finite rate of increase) in an ordinary (nondisastrous) season. By this measure, a genotypic population which increased by, say, 50% each season would be said to have a fitness of 1.5.

The fitness computations are straightforward. Because the population is large, it is to be expected that in each season, about 3/8 of the ground nesters will lose their offspring to predators while the remainder will raise two offspring each. If the ground-nesting subpopulation is of size N at the beginning of a season, it will therefore attain a size of $(3/8)N \times 0 + (5/8)N \times 2 = 1.25N$ by season's end (i.e., the ground-nesting genotype has fitness 1.25). The fitness of the tree-nesting genotype, on the other hand, is exactly 1.0. The ground nesters are therefore fitter than the tree nesters, so that the latter genotype, though occasionally appearing through mutation, cannot be maintained. By standard evolutionary reasoning, then, natural selection would favor the ground-nesting instinct. A biologist would predict that the ground-nesting behavior would probably be the one found in nature under the circumstances of the example.

The Decision-theoretic Reinterpretation

The essential background facts of the foregoing problem situation could have been presented diagrammatically as shown in Figure 2.1. The lines emanating from the square node at the left end of the diagram represent the possible behaviors under comparison. The upper line, associated with the ground-nesting subpopulation, branches in two at a round node standing for the possibility of predation, the fractions of this subpopulation experiencing predation or nonpredation of their offspring being shown over the upper and lower branches, respectively. The numbers of successfully raised offspring per individual are shown after the extreme right ends of the paths (solid circles)—for example, a ground nester experiencing no predation will raise 2 offspring (end of middle path).

The fitness computations required to solve the problem could also have been organized with the help of such a diagram. This is shown in Figure 2.2, which augments the data of Figure 2.1 with the fitness results. First, the fitness of each of the three ultimate subpopulations is written over each corresponding path's end; it is simply the number of offspring per individual already indicated to the right of each solid circle. Then, for the hollow round node, the fitness of the corresponding genotype is calculated by summing the products of the numbers over the paths leading rightward out of it (i.e., by computing the weighted average $(3/8) \times 0 + (5/8) \times 2$ and writing the result over the node). The two fitnesses of ultimate interest now appear above the nodes to the immediate right of the square node. Since the upper one shows the greater fitness, the lower path has a double line across it to indicate that evolutionary reasoning makes the corresponding behavior, tree nesting, biologically implausible.

The same diagrammatic procedure could conveniently be used to handle more complex problems involving more than two competing behaviors and more than one kind of environmental risk. The generalization is straightforward. Briefly, after constructing the appropriate diagram, one calculates weighted averages for the round nodes by working from right to left until they all have a fitness written over them. The path from the leftmost (square) node to the neighboring (round) node with the largest associated fitness indicates the behavior most likely to exist in nature.

Now, there is nothing biologically novel in any of this, and any evolutionist could have used similar diagrams at any time as a heuristic device for visualizing the form of standard biological reasoning. However, it cannot have escaped the notice of readers familiar with decision theory that such diagrams are very like the "decision flow diagrams" often drawn by decision theorists (the particular conventions adopted here are those of Raiffa 1968). In fact, by taking a few liberties, the diagrams can even be given a suggestive decision-theoretic interpretation. One imagines the individual organism to be the reasoner or "decisionmaker," and the diagram to be a device for deducing which of the possible acts available to it would be the most advantageous or "rational." The root (square) node is the "decision node," and the paths leading out of it represent the possible alternative "acts." The hollow round nodes are "chance nodes" standing for events. These events can happen with the probabilities indicated over the paths leading rightward out of them. The numbers at the branchtips are the possible ultimate "consequences" that the decisionmaker must consider. Each of these has a certain "utility" to him, namely, the number written above the branchtip. The numbers above the chance nodes are "expected utilities" calculated as probability-weighted averages in the obvious way. The decision is, of course, made by maximizing expected utility. In the present example, the "rational" decision is to nest on the ground. All paths out of the root node other than the maximizing one are crossed with a double line to indicate that the corresponding acts are blocked off to a rational decisionmaker.

Clearly, there is a close formal analogy between a conventional biological analysis which seeks to predict or explain organismic behavior, and traditional decision-theoretic methods intended to characterize rational behavior. Biologists assert that organisms, humans presumably included, tend to possess those behaviors which maximize their fitness. Decision theorists assert that decisionmakers do or should choose those acts that maximize their expected utility. Evidently, there is a deep-rooted parallelism between the analytic frameworks of the two disciplines.

Extending the Comparison

Before attempting to assess the significance of this parallelism, it may be helpful to gain some feeling for how far the correspondence extends. The example of the ground and tree nesters is, of course, trivial and artificial, but it can be changed, extended, elaborated, and generalized in a number of directions until most of the more commonly considered biological complications have been taken into account. For many of these extensions, the parallel with classical decision theory is strictly preserved; for others, it is not. A few of the possible extensions have been explored elsewhere (Cooper 1981); here, we will content ourselves with a brief and tentative summary.

Among the generalizations of our example for which classical decision computations remain biologically valid are the following. The example unrealistically assumed an invariable number of offspring for each set of circumstances (e.g., precisely two surviving offspring for each unmolested ground nester). This artificiality can be removed by generalizing from a specific number to a probability distribution, and when this is done, the biological/decision-theoretic parallelism survives intact. The example introduced only one kind of chance event (possible predation) which must have one of only two possible outcomes, but generalization to any finite number of random environmental variables with any number of possible outcome values is also straightforward. As already mentioned, more than two potential behavior patterns may be compared, leading in the extreme case of a continuum of possible behaviors to a theory of quantitative choice. More fundamentally, the decision node need not always be at the root of the tree, and there can be more than one decision node. This generalization allows for conditional strategies (e.g., "If no predators have been sighted lately, nest on the ground"), and primitive learning can also be represented

("If a predator is seen three or more times in an area, avoid that area thereafter").

Elaborations of the example which can *not* be analyzed adequately using only simple classical decision computations include generalizations to temporally changing ("coarse-grained") environments and to organismic lifetimes containing more than one breeding season (iteroparity). A different choice of fitness measure can also necessitate a departure from ordinary decision-theoretic computational patterns, as can an assumption of sexual reproduction. In such circumstances, classical decision calculations would be biologically inadequate, and a procrustean application of them could produce faulty predictions. It is perhaps worth remarking, though, that classical computations, even when they constitute an oversimplification of the biologically correct analysis, can often be applied in such a way that they tend usually to produce the same "decisions," and in that sense, may enjoy a certain approximate validity.

Three Reductionist Hypotheses

There are at least three ways to interpret the assertion that decision theory is reducible to evolutionary biology. The first is not directly germane to the question of reductionism, but will be mentioned anyway because it might otherwise be mistakenly thought to be what is at issue. It is the assertion that *evolutionary theory predicts that humans and other organisms should tend to behave in a way which is largely consistent with the norms of decision theory.* This claim is plausible enough, for rationality is clearly a generally fit quality (c.f. Sober 1981). Indeed, if decision theory delineates something approximating "rational" behavior, and rational behavior is generally adaptive, and evolutionary theory predicts that what is most adaptive is also most likely to be found in nature, then it follows immediately that there should be a tendency for decision theoretically prescribed behavior to appear in nature. As Shepard has remarked (1964), "The rules that govern the decisions made by actual organisms have been selected precisely on the basis of their long-range contribution to survival and propagation."

The problem with this vague version of reductionism is not that it is mistaken, but that it does not say enough to be interesting. In particular, it does not say anything about what causes the rules of decision to take the particular mathematical forms they do. They could, for instance, have their source in some platonic mathematical heaven entirely unrelated to biology, and the claim could still be true. Thus, virtually any decision theorist who accepts Darwinism at all, or any biologist who sees virtue in decision theory at all, could probably accept some version of such a thesis without further ado. It does not even specify clearly which science is supposed to be reducible to which. We therefore set it aside as too loose to merit further discussion.

A more interesting hypothesis is that *the formal rules of descriptive decision theory are directly derivable from the laws of evolutionary theory,* or in a slightly stronger form of the hypothesis, perhaps even identical with some of those laws. According to this thesis, the true underpinnings of descriptive decision theory are the fitness-maximization techniques employed by biologists to predict or explain organismic behavior. If this is true, decision-theoretic rules should prove descriptively accurate to the same extent that the evolutionary generalizations from which they are drawn are scientifically well-founded. The boundary line between descriptive decision theory and ethology is an arbitrary one, since the biologist, upon commencing an evolutionary optimization study of individual choice behavior, immediately becomes a descriptive decision theorist as well. We shall call this the *Descriptive Reduction Hypothesis.* Note that it says nothing about normative correctness. In particular, it leaves open the possibility that evolutionary theory might yield rules of choice behavior which are descriptively accurate but nonetheless mistaken or irrational when compared against some absolute standard of consistency or coherence.

The third and strongest hypothesis is what we shall call the *Normative Reduction Hypothesis.* It states that *the formal rules of normative decision theory are directly derivable from the laws of evolutionary biology.* By this hypothesis, evolutionary considerations actually define what rational behavior is: in advising decisionmakers how they should act, the decision theorist ought to consult evolutionary theory. Obviously, this is apt to be the most controversial of the hypotheses.

Discussion of the Descriptive Reduction Hypothesis

Modern neo-Darwinist evolutionary biology is a vast science which, among other things, facilitates the scientific prediction and explanation of organismic characteristics. Included among them are behavioral properties. "Regularities of behavior are as predictable and discernible as the color of plumage and the shape of eggs" (Tiger and Fox 1971, p. 17). To an adherent of the Descriptive Reduction Hypothesis, what twentieth-century decision theorists have done is, in effect, to reconstruct (partially, independently, and in apparent ignorance of the relevant biology) just that fragment of elementary evolutionary theory which bears most directly on the prediction of general behavioral properties. While this is a remarkable achievement and a dramatic convergence of two sciences, the reductionist suspects that still more might be accomplished if the underlying biological basis of decision theory were recognized explicity.

Of special interest in this connection is the fact that, while the simplest evolutionary models imply a descriptive theory of decision which coincides with garden-variety decision theory, more refined evolutionary analyses often lead (as mentioned earlier) to unexpected elaborations of the usual classical theory, of which the latter is, at best, a simplistic first approximation. Moreover, descriptive decision theorists have found that, in general, the classical decision theory is not an especially good predictor of actual human choice behavior: it is often necessary to invent ad hoc psychological explanations of the discrepancies between what is classically predicted and what is observed. Might some of the discrepancies be accounted for if certain biologically indicated elaborations of the classical theory were made? The decision theorist who subscribes to the Descriptive Reduction Hypothesis sees evolutionary biology as a potential source of possible modifications of the classical decision rules, modifications which might make them at the same time more biologically realistic and more descriptively accurate.

The imperfection of the formal parallelism between classical decision models and sophisticated evolutionary models may help to explain a point which has puzzled some commentators. Reijnders writes, "It is the author's opinion that viewed in the light of the assumption that natural selection will favor intuitive statisticians implicit in the theories of Lewontin (1961) and Maynard-Smith and Price (1973), the phenomenon that man is apparently not an intuitive statistician . . . comes as a surprise" (1978, p. 246). Reijnders refers here to the fact that humans have been found in psychological experiments to be rather sloppy intuitive statisticians when judged against classical decision-theoretic norms. But humans might turn out to possess considerably more impressive statistical intuitions if their behavior were analyzed in the light of decision rules derived from more refined biological theories. If so, their apparently mediocre performance in experiments to date would be seen in retrospect as a mere artifact of the unwitting use of an oversimplified evolutionary model.

It has been suggested elsewhere that the activity of deriving decision rules from evolutionary models and comparing them with observed behavior might appropriately be called *Natural Decision Theory* (Cooper 1981). Natural decision theory is essentially the study of decision rules viewed as a product of natural selection and possibly other natural evolutionary forces. It is descriptive decision theory, but on a biological foundation. Such a theory of choice behavior can be carried to any desired degree of refinement, depending upon the degree of biological detail introduced into the underlying evolutionary representations. Classical decision theory then becomes that special case of natural decision theory which is obtained by confining attention to a certain class of especially simple evolutionary models (Cooper, in press).

In illustration of the concerns of natural decision theory, an intriguing preliminary result may be cited. When the classical expected utility theory is generalized in such a way as to accommodate evolutionary models involving temporally changing environments, the possibility immediately arises that mixed strategies may be advantageous (Cooper and Kaplan 1982; Kaplan and Cooper 1984). That is, it may sometimes be less fit for an organism to follow its optimal pure strategy than to randomize its choices by doing something equivalent to flipping a coin or rolling a die. This is, of course, a radical departure from classical decision theory, according to which such randomization is supposed never to be advantageous in a non-game-theoretic situation. If the analysis is correct, it may well be a common occurrence for real organisms, possibly including humans, to violate classical decision criteria by strategy mixing. This effect alone could help to account for much seemingly irrational decision behavior (preference intransitivities, etc.) observed by experimenters.

Objections to the Descriptive Reduction Hypothesis

An objection to the Descriptive Reduction Hypothesis which is apt to occur immediately to a traditional decision theorist is that it associates utility with fitness, and that this association is unjustified. The relationship of utility to fitness is indeed problematic, and is unlikely to be resolved fully by anything that could be said here. However, it may be worth remarking that classical decision theory has accustomed all who have studied it to a concept of utility which is, so far as can be seen from the axioms alone, empirically content-free, asserting nothing substantive about what utility actually is except that an ideal reasoner's utility measure must satisfy certain logical and mathematical constraints. Naturally, any proposal to inject actual scientific substance into the utility notion then comes as something of a shock. But the shock itself does nothing to refute the reductionist thesis. It might argue against the converse—that biology is derivable from decision theory—but, of course, this is not what the hypothesis asserts. If utility is to be assigned any specific scientific interpretation at all, then that interpretation must necessarily be narrower than the classical characterization, and we must be prepared for the specificity.

It should also be recognized that fitness is a far more encompassing notion than is generally appreciated outside the biological community. In recent decades, a staggering variety of behavioral phenomena ranging from foraging habits to altruism have been scientifically explained in terms of fitness maximization. Virtually all of sociobiology consists of a drawing out of the consequences of the fitness concept, and the recent surge of interest in that subject is at least partly due to an increased scientific ability to explain human and animal behavior in terms of adaptive fitness (Wilson 1975; Caplan 1978). Few who are familiar with this literature would be willing to dismiss as preposterous the idea that human and animal decision criteria might turn out to be fitness-related at even the highest levels of abstraction.

Another objection likely to be raised against the Descriptive Reduction Hypothesis is that decision theory is supposed to be about decisionmaking, whereas in the simplest biological analogues, it is often questionable whether any genuine decisionmaking is taking place. Thus, in our earlier ground-nesting/tree-nesting example, there is nothing resembling an organism that is at first unsure which act to perform, that puzzles over it, applies decision criteria to it, and eventually becomes sure of what to do. In fact, there is no judgment, no "cogitation" at all by the alleged "decisionmaker"; instead, one has only an external scientific observer deducing which genetically preprogrammed instinct is likeliest to prevail.

The objection is a natural one, but is probably insufficiently imaginative in seeing how such examples can be extended to more interesting cases, and at the same time, insufficiently sensitive to the degree to which all so-called choice judgments are open to the charge of determinism. If a gene for a fixed simple instinct can be postulated, it is but a step to imagining a gene for a behavior whose character is conditional upon prior environmental input, and another from there to a gene or gene complex for learning and finally for "cogitating" a prospective choice. An ideally adapted cogitation mechanism would, of course, be one which tended to reproduce the results of a sound evolutionary fitness-maximization analysis. The cogitation would then be an evolved internalization of a capability for fitness analysis, carried out not by a scientific observer, but by the individual organism itself. In this perspective, it is hardly surprising that the decision rules arrived at by nonbiologist philosophers have resembled the rules of a simplified biological fitness analysis.

A technical hurdle confronting the reductionist program is that although the central importance of fitness is universally conceded among evolutionists, there is as yet no consensus as to the most appropriate way to measure it mathematically (Stearns 1981; Mills and Beatty 1979;

Rosenberg 1982). Perhaps different measures are appropriate under different circumstances, with all logically defensible measures of adaptedness being in some sense derivative from some single, logically fundamental underlying measure (Cooper 1984).

Finally, it could be objected that current evolutionary theory is simply not powerful enough yet to support detailed explanations of all the intricacies of human and animal choicemaking. Natural selection is the chief—but by no means the only—force at work in the evolutionary process. Even natural selection is not yet fully understood in all its ramifications, and the other forces have probably not yet even been fully identified (c.f. Gould and Lewontin 1979). This objection seems well taken. In view of it, the Descriptive Reduction Hypothesis should for the foreseeable future probably be interpreted as a working assumption to the effect that descriptive decision theory is *in part* derivable from known evolutionary theory. Natural Decision Theory then becomes the activity of looking for biologically well-founded descriptive decision rules, to whatever extent these can be derived from current evolutionary knowledge.

Discussion of the Normative Reduction Hypothesis

That decision rules derived solely from evolutionary biology could ever have *prescriptive* value will seem preposterous to anyone who is a logical absolutist—who denies that logical laws can be dependent upon contingent scientific generalities—and who further believes that decision theoretic rules enjoy the status of logical laws. Nevertheless, it is conceivable that in the inevitable confrontation between logical absolutism and reductionism in decision theory, it is the reductionist view that will ultimately prevail.

Those who defend the normative value of classical decision rules claim not that the rules are, but that they *should be,* followed. However, the term "should" is used in a peculiar sense. It is not asserted that it would be ethically wrong to disobey these rules; rather, what is claimed is that one had better follow them if one wishes to act in accord with one's own ultimate aims. Indeed, among the alternative names for utility suggested by early commentators are "desiredness" (Viner 1925) and "wantability" (Fisher 1918). More recently, it has been suggested facetiously that utility might be measured with a "hedometer" (Jeffrey 1981). Evidently, then, the "normative" element in decision theory boils down to a matter of clarifying ultimate individual wants. It is hard to see why pure logic should have a monopoly in this matter, or why it should be assumed that biology has nothing to contribute.

It will be objected that the role of prescriptive decision theory is merely to ensure consistency and coherence when acting upon one's desires, and that these, at least, are purely logical matters. But this raises the awkward question of why we should want our behavior to be consistent in the exact sense of the Bayesian expected utility formula, or why we should always scrupulously avoid Dutch book bets. We are, after all, biological creatures. If it should turn out that other non-Bayesian consistency criteria are sometimes biologically fitter (and so probably more congenial to our ultimate aims and desires), why should they be considered any less normative? Such questions seem foolish only until one's initial classical prejudices have had a chance to subside.

Classicism Reconsidered

Suppose for discussion's sake that we were to accept both reduction hypotheses, radical as they may seem. Would it necessarily follow that we must reject classical decision theory altogether? As noted earlier, the usual classical decision rules are in apparent conflict with those derivable from realistic evolutionary models. Does this mean that the classical rules are just plain illogical, and that the luminaries who gave us our modern theory of rational choice were simply mistaken? No, not necessarily. Another view, slightly more conservative but still in accord with the reduction hypotheses, is available. It is that the classical theory is not logically mistaken, but has so far been universally mis-applied. Granted, the misapplication has not been so serious as to render the theory totally unviable, but it is a misapplication nonetheless.

In this view, the classical theory has been applied in an oversim-plified way through a failure to distinguish appropriately among different evolutionary levels. Specifically, the oversimplification has lain in the assumption that the decision-making agent is the individual organism acting directly for its own individual benefit. This assumption is a biological error. As twentieth-century evolutionists have come to recognize, adaptive forces will normally cause an individual to wish to act as though for the benefit of its genotype or some similar higher level entity. In biological jargon, the individual would be said to exhibit a phenotypic behavioral tendency to maximize the fitness of the higher evolutionary unit. (Dawkins 1976, c.f. 1982, has provided an excellent exposition of this point for the nonbiologist.) Now if this be so, a correct application of classical decision theory—one which maximizes the expected utility of the appropriate unit—should deal directly with the probabilities and utilities that affect the higher level entity whose fitness is at issue. These are not necessarily the same as the probabilities and utilities of the events and consequences encountered by the individual organism, nor are the kinds of computations necessarily the same. In general, a careful translation will be needed before it can be seen how the higher level considerations bear on those of the individual. The misapplication of classical decision theory has consisted in the omission of this translation step.

What emerges is a two-tier conception of how decision theory ought to be applied. On the upper tier, classical decision theory maximizes the expected utility (fitness) of the genotype or other relevant higher level evolutionary unit. This is the "gene's-eye view" of which evolutionists sometimes speak. Such a maximization, when translated onto the lower or individual level, takes the form of a set of decision rules relevant to the circumstances of the individual organism—of what each particular individual should do if it experiences such-and-such a local event. To be sure, the individual is still maximizing an expected utility by the classical rules, but it is a higher level fitness and not his own individual subjective utility in the sense to which decision theorists have become accustomed. That is why his rules may look nonclassical when examined without benefit of the gene's-eye view. Of course, in sufficiently simple evolutionary models, the form of the lower level rules closely resembles the form of the classical higher level ones. But when more complex evolutionary models are introduced, the lower level rules become more involved; hence, the need for a "natural" decision theory to interpret the classical on the individual level.

This split-level view maintains the integrity of the classical theory of decision, and at the same time, reconciles it with the development of a natural decision theory applicable to individual decisions. It takes into proper account the biological complexities of fit choicemaking, and explains how a theory embracing these complexities could have more descriptive and explanatory power than the classical theory as usually applied without benefit of biology. It also clarifies the sense in which even a classicist might regard a nonclassical, "natural" decision theory as normative for the individual. It is not the only possible philosophy, especially since the discovery that the classical theory has no absolute validity on the individual level might prompt one to wish to reexamine the question of whether it has absolute validity on any level. But the two-tier perspective does offer at least one plausible conceptual platform from which the reductionist hypotheses can be viewed without alarm.

Summary

In sufficiently simple biological circumstances, the mathematical methods of evolutionary optimization analysis look curiously like the rules of classical decision theory. Under more complex conditions, on the other hand, standard evolutionary techniques depart somewhat from classical decision rules, but might reasonably be expected to come closer to describing actual decision behavior. In this situation, one is naturally led to ask whether there might be a reducibility relationship linking decision theory to evolutionary biology. If so, the possibility arises of exploiting evolutionary optimization methods as a source of plausible hypotheses for descriptive decision theory; and more controversially, perhaps even for normative decision-theoretic purposes. In any case, such a reduction would be of more than ordinary interest because it would reinterpret what many have regarded as a foundational logical calculus as a branch of an empirical science.

Earlier, it was indicated informally by means of an example how the classical decision rules might be derived from a simplified evolutionary model. It remains to carry out the derivation formally, and to investigate in more detail the ways in which elaborations of the model lead to nonclassical "natural" decision rules. It also remains to be verified experimentally that such biologically derived rules do, in fact, tend to be more descriptively accurate than classical ones. Another question calling for careful examination is the extent to which the reducibility of decision theory—assuming it is indeed reducible—carries over into game theory. Finally, the philosophical issue of whether natural decision theory might be normative for the individual calls for open-minded consideration.

Bibliography

Caplan, A.L. (ed.), 1978: *The Sociobiology Debate—Readings on the Ethical and Scientific Issues Concerning Sociobiology*, Harper and Row, New York.

Cooper, W.S., 1981: Natural decision theory—a general formalism for the analysis of evolved characteristics, *Journal of Theoretical Biology* 92:401–415.

Cooper, W.S., 1984: Expected time to extinction and the concept of fundamental fitness, *Journal of Theoretical Biology* 107:603–629.

Cooper, W.S., 1987: Decision theory as a branch of evolutionary theory—a biological derivation of the Savage axioms, *Psychological Review* (in press).

Cooper, W.S., and R.H. Kaplan, 1982: Adaptive "coin-flipping"—a decision-theoretic examination of natural selection for random individual variation, *Journal of Theoretical Biology* 94:135–151.

Dawkins, R., 1976: *The Selfish Gene*, Oxford University Press, Oxford, England.

Dawkins, R., 1982: *The Extended Phenotype—The Gene as the Unit of Selection*, W.H. Freeman, Oxford, England.

Edwards, W., 1954: The theory of decision making, *Psychological Bulletin* 51:380–417. Reprinted in W. Edwards and A. Tversky (eds.), 1967: *Decision Making*, Penguin Books, Baltimore, Md.

Einhorn, H.J., and R.M. Hogarth, 1981: Behavioral decision theory—processes of judgment and choice, *Annual Review of Psychology* 32:53–88.

Fishburn, P.C., 1981: Subjective expected utility—review of normative theories, *Theory and Decision* 13:139–199.

Fisher, I., 1918: Is "utility" the most suitable term for the concept it is used to denote?, *American Economic Review* 8:335–337. Reprinted in A. N. Page (ed.), *Utility Theory—A Book of Readings,* Wiley, New York, 1968.

Gould, S.J., and R.C. Lewontin, 1979: The Spandrels of San Marco and the Panglossian paradigm—a critique of the adaptationist programme, in J. Maynard Smith and R. Holliday (eds.), *The Evolution of Adaptation by Natural Selection,* The Royal Society, London.

Jeffrey, R.C., 1981: The logic of decision defended, *Synthese* 48:473–492.

Jeffrey, R.C., 1983: *The Logic of Decision* (2nd ed.), McGraw-Hill, New York.

Kaplan, R.H., and W.S. Cooper, 1984: The evolution of developmental plasticity in reproductive characteristics—an application of the "adaptive coin-flipping" principle, *American Naturalist* 123: 393–410.

Kyburg, H.E., and H.E. Smokler (eds.), 1980: *Studies in Subjective Probability* (2nd ed.), R.E. Krieger Publishing Co., Huntington, N.Y.

Lee, W., 1971: *Decision Theory and Human Behavior,* Wiley, New York.

Lewontin, R.C., 1961: Evolution and the theory of games, *Journal of Theoretical Biology* 1:382–403.

Maynard-Smith, J., 1978: Optimization theory in evolution, *Annual Review of Ecology and Systematics* 9:31–56.

Maynard-Smith, J., and G.R. Price, 1973: The logic of animal conflict, *Nature* 246:15–18.

Mills, S.K., and J.H. Beatty, 1979: The propensity interpretation of fitness, *Philosophy of Science* 46:263–286.

Nagel, E., 1961: *The Structure of Science,* Harcourt, Brace and World, New York.

Raiffa, H., 1968: *Decision Analysis: Introductory Lectures on Choices Under Uncertainty,* Addison-Wesley, Reading, Mass.

Ramsey, F.P., 1926: Truth and probability, in R.B. Brathwaite (ed.), 1950: *The Foundations of Mathematics and Other Logical Essays,* The Humanities Press, New York. Reprinted in Kyburg and Smokler (1980, op. cit.).

Reijnders, L., 1978: On the applicability of game theory to evolution, *Journal of Theoretical Biology* 75:245–247.

Rosenberg, A., 1982: On the propensity definition of fitness, *Philosophy of Science* 49:268–273.

Savage, L., 1972: *The Foundations of Statistics* (2nd ed.), Dover, N.Y.

Schaffner, K.F., 1977: Reduction, reductionism, values, and progress in the biomedical sciences, in R.G. Colodny (ed.), *Logic, Laws, and Life,* University of Pittsburgh Press, Pittsburgh, Pa., 143–171.

Shepard, R.N., 1964: On subjectively optimum selections among multi-attribute alternatives, reprinted in W. Edwards and A. Tversky (eds.), 1967: *Decision Making—Selected Readings,* Penguin Books, Baltimore, Md.

Slovic, P., B. Fischhoff, and S. Lichtenstein, 1977: Behavioral decision theory, *Annual Review of Psychology* 28:1–39.

Sober, E., 1981: The evolution of rationality, *Synthese* 46:95–120.

Stearns, S.C., 1981: On fitness, in G. Roth (ed.), *Fourth Bremen Symposium on Biological Systems Theory,* Bremen, Germany.

Templeton, A.R., and E.D. Rothman, 1974: Evolution in heterogeneous environments, *American Naturalist* 108:409–428.

Templeton, A.R., and E.D. Rothman, 1978: Evolution and fine-grained environmental runs, in C.A. Hooker, J.J. Leach, and E.F. McClennen (eds.), *Foundations and Applications of Decision Theory* (Vol. 2), D. Reidel, Dordrecht, Holland.

Tiger, L., and R. Fox, 1971: *The Imperial Animal,* Holt, Rinehart and Winston, New York. Excerpted in Caplan (ed.), op. cit.

Viner, J., 1925: The utility concept in value theory and its critics, *The Journal of Political Economy* 33:638–659. Reprinted in A.N. Page (ed.), 1968: *Utility Theory—A Book of Readings,* Wiley, New York.

von Neumann, J., and O. Morgenstern, 1953: *Theory of Games and Economic Behavior* (3rd ed.), Princeton University Press, Princeton, N.J.

Wilson, E.O., 1975: *Sociobiology—The New Synthesis,* Harvard University Press, Cambridge, Mass.

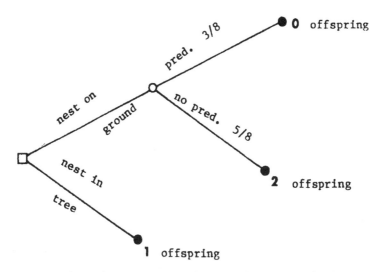

FIGURE 2.1 A decision tree diagram for an evolutionary
decision situation in which an organism must "decide"
whether it is fitter to nest on the ground or in a tree.
Nesting on the ground exposes an individual to predation
with probability 3/8. By convention, square nodes represent decision nodes, hollow round nodes are chance nodes,
and solid round nodes are consequence nodes.

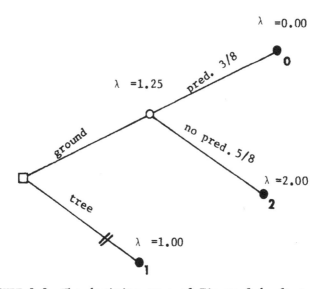

FIGURE 2.2 The decision tree of Figure 2.1 after application of the decision procedure. The values above the
round nodes are fitnesses. The double slash across the
lower branch indicates that a decision to nest in trees
is not the fittest possible.

3. Evolutionary Computing by Dynamics in Living Organisms

Introduction

In this polemical chapter, I argue that computer metaphors and information science can be inimical to clear thinking about evolutionary problem-solving processes in biology, whether at the organism level or at the species level. The limitations of the computer metaphor and symbol strings are briefly examined, and contrasted with the hardware changeability of biological systems and the essential role of structural change in biological "problem solving." In biology, on any time scale, dynamical features of the physical situation dominate. Two examples are given: geotropism in plants and the detection of odors by rabbits. Information processing in the technological world of silicon and in the biological world of carbon are contrasted—a contrast that is heightened by comparing a view of the universe as an information-driven, finite-state automaton, with a view of the universe as a dynamical system breaking symmetries and creating new forms and functions as it evolves.

My aim is to show that a purely "informational" view of computation, no matter how elaborated, fails in principle to account for the evolutionary, adaptive, self-organizing, problem-solving properties of biological systems. It fails to account for either the ontogenesis and behavior of a particular organism, or the phylogenesis of a species over longer times.

Different Meanings of the Term "Evolution"

In mathematics, "evolution" sometimes refers to the extraction of a root from any given power; in engineering, it frequently refers to the following of a bounded trajectory from initial conditions to final equilibrium or steady state according to a dynamical rule or law under constraints; in physics, broadly, it is the overall process in which the whole universe can be seen as a progression of interrelated phenomena (starting, according to current wisdom, with the "Big Bang"); in biology, it is the historical development of a group, as well as the theory that various types of animals and plants have their origin in other pre-existing types, and that the distinguishable differences are due to modifications in successive generations.

In computer science, the term "evolution" has meanings such as may apply to learning circuits (Conrad 1974; Hopfield 1982; Waner and Hastings 1987). In these instances, an "evolutionary learning system" incorporates both a state space whose individual states are modes of information processing, and an "evolutionary" search through these states which follows simulated annealing dynamics. The notion of annealing dynamics has been clearly explained by Kirkpatrick et al. (1983) and by Hinton et al. (1984).

The focus of this chapter will be on changes in dynamic trajectories of living organisms as these are interpreted by us (with deliberate anthropomorphism—often conveniently forgotten) as being "adaptive." In order to avoid confusions between the terms "adaptive behavior" and "learning" (as may arise in discussions of changes in animal behavior seen in classical avoidance conditioning paradigms, or in appetitive conditioning paradigms), I shall begin the discussion with analysis of selected behavior of plants. Conveniently for my arguments, they have no nervous systems with which to learn. Then, I shall address the case of how an animal learns to recognize a smell, and adapt its behavior to it. A detailed account of adaptability, more general than that I shall invoke, has been provided by Conrad (1983).

Without denying that new developments in evolutionary computing (e.g., "annealing" searches or "genetic" breeding and selection processes leading to automata optimized for specific tasks) show signs of helping to solve problems previously difficult to handle, I shall argue that no discrete-mode, finite-state automaton can support a process that is evolutionary in the biological sense of adaptive behavior (and most certainly not those processes involved in the evolution of the biosphere itself).

Complaints About Internal Representations and the Equating of Information with Symbol Strings

In several papers, Kugler and Turvey and their colleagues have called attention to the inadequacies of the computer metaphor as a basis for accounting for complex, self-organizing (especially living) systems (Turvey and Kugler 1984; Carello et al. 1984). Symbol strings can adequately account for the classical indicational and injunctional aspects of information. These categories include the Shannon-Weaver information of the selective kind, and also much descriptive information (e.g., resolution of a measurement). Selective information estimates, among other things, how surprised we would be if a symbol or symbol string occurred twice in succession; injunctional information directs or commands states of affairs ("stop!"). In these cases, the bit, thought of as occupying a two-state physical degree of freedom as in the setting of a switch, is an element of stored information.

Kugler and Turvey emphasize that indicational and injunctional information requires also a companion specificational sense of information (a concept introduced by Gibson that is distinct from the other orthodox senses of the term information). Gibsonian specificational information is a physical variable that can be associated with low-dimensional macroscopic properties of low-energy physical fields (i.e., kinematic fields) lawfully generated by properties of system and surround. Optical flow fields provide a convenient example.

One can know to stop a car because he or she can read a symbol string of the type "SLOW DOWN" or "STOP" that is intended to direct the dynamics of traffic flow. But even so, complying with these injunctions is possible only if there is a continuously available information flow specific to the retarding of forward motion, and the time to contact with the place where the velocity is to go to zero. Both the deceleration of global optical flow (the pseudomotion of fixed objects in the surround as one moves through a textured scene) and the rate of dilation of the visual solid angle to the point of observation that is created by the approach to the place where the motion is to be fully arrested together specify continuously the time at which the place will be contacted (Turvey and Kugler 1984; Yates and Kugler 1984). Without information in the specificational sense, information in the indicational/injunctional sense is impotent.

Specificational information arises from the physics of the situation: symbol strings are not involved, nor are internal representations. The optical flow field, with the significant singularity in it, is self-organizing out of the motion. An education of attention (Gibson's phrase) is required to attach significance to the singularity, but no pre-existing internal representation is required to allow the nervous system to assess impending contact with a fixed object, such as a stopped car ahead. (These points have been explained in considerable detail in the references cited, and will not be further elaborated here.)

The notion of an internal "representation" of "external" reality, whether it appears in the field of artificial intelligence or psychology, can unsettle clear thinking about adaptive behavior in living systems. The existence of such representations is unprovable, and in any case, the concept is unnecessary. As Freeman and Skarda (1985) remark (while disagreeing with the suppositions involved, as I do, too):

> Commonly it is said that a stimulus is encoded in or stored as or represented by a pattern of (neural) activity, as though a certain token or symbol or message were introduced into the brain along with a certain meaning. Likewise it is postulated that in toto the brain sustains a world view, image, model or representation of the environment, that is continually updated by new input, and that provides the basis for patterned output or action. Thus recognition is conceived as based on comparison of a stimulus-evoked activity pattern with each of a store of pre-existing patterns, and decision as a selection from a store of motor patterns of one that is optimal according to certain rules that are also encoded and exercised perhaps in the manner of a binary decision tree. Actions can result from representations serving centrally in the manner that stimuli do peripherally.

Freeman and Skarda go on to deplore this concept, and note that the root of the difficulties is that the referents, contexts, and meanings of representations are invariably in the brain of the observer and not that of the observed. By comparing the behavior of animals responding to familiar smells with the behavior when responding to novel smells, they show that the animals (in this case, rabbits) react to novel odors in very short time spans that are inconsistent with the time required for an exhaustive review of all the odors that we suppose adult rabbits to have learned to identify, as indeed they must in order to survive in an ever-changing environment. Representation by a "search image" would require that some pattern pre-exist the stimulus. If the pattern were ascribed to one of a collection of synaptic "templates" created in the (olfactory) bulb by learning, then odor identification would require such a systematic search through the collection. If the odor is novel, the entire collection would have to be reviewed, contrary to experimental results on the times to response.

Repeatedly, we see that the language of representation involving feature extractions, information encoding in patterns of neural activity, formation of search images, tests of hypotheses based on learning, and other seductive phrases turns out to be empty rhetoric that leads nowhere. To quote Freeman and Skarda:

> They are not merely wrong; they are irrelevant and misleading. It is our conclusion that animals interact with their environments but need not and do not represent them in doing so. An ultimate goal in neurophysiology is to explain human consciousness and our capacity for rep-

resentation, but experiments on animal brains will not directly support that, and the concept should not be part of its explanation.

The paper by Freeman and Skarda is so rich in experimental results, statistical methods, careful experimental designs (executed in four paradigms over a 12-year period), and critical commentary that I can't do it justice here. I can only emphasize that the reader who has been pressed to believe that computers reveal some essence of brains (other than that brains design them) may find relief in the account by Freeman and Skarda.

Brains Versus Machines

Elsewhere, I have examined the relationship between computation in the world of silicon (computer science) and computationlike processes in the world of carbon (biology) (Yates 1984). That essay addressed six questions that had been debated for five days during a conference on "Chemically-based Computer Designs" organized by Michael Conrad and me:

1. Are there fundamental, quantum mechanical limitations on computation? (Answer: No)
2. Are there fundamental thermodynamic limits on computation? (Answer: No)
3. Are there fundamental limits to problem solving by formal (Church-Turing) computation implemented by digital computers as finite state machines based on networks or binary switches? (Answer: Undecided)
4. What are the practical physical limitations on computer design? (Answer: Discussions here involved exponential time and exponential resources problems, heat dissipation, quantum mechanical tunneling . . . but if these are practical limitations that may be reached when we try to extend current practices of digital computation toward greater miniaturization of switching circuits, we are nevertheless still very far from reaching those limits.)
5. What are the potential contributions of molecular electronics (especially carbon-based) to digital computer design? (Answer: More work needs to be done on soliton propagation in molecular networks, and on the properties of microtubules as microcomputers)
6. Do biological systems inspire technological imitations for the purposes of computer design? Should we imitate, with other materials and processes, any biological functions to design new

computers? Should we attempt to use biological functions or materials directly to design chemically based computers? (The answers to these questions were extensive and varied, and will not be summarized here.)

The machine-mind problem came up in various guises, and I will summarize some of the issues to complete the background for the main arguments of this chapter, namely,

1. that in biological systems, information either as structure or as singularities in kinematic flow fields acts to constrain kinetic processes,
2. that dynamics was prior to information, and generated it in the history of the biosphere, and
3. in the history of an organism, biological evolutionary "computing" is chiefly dynamics, contrary to technological computing, which is chiefly information.

The article by Churchland, "Is Thinker a Natural Kind?" (1982), highlights the conflict between those who believe that the human mind is a program that (in principle) could run on non-neural machines, and those who believe that if it is a program at all, it can run only on a neural substrate. Earlier, the same issue was considered by Watanabe, who examined four positions on the subject (1974):

The mind is not a formal system, and although one might be able to program some superficial aspects of behavior, the really important and significant aspects of intelligence—the ones that stand out as "human"—cannot be captured in a formal system.

(Watanabe uses the term "formal system" loosely to denote something capable of exact specification, i.e., something that, in principle, could be simulated by a Turing machine or by a program on a digital computer.)

The mind is not a formal system, but many of its activities may be simulated by the digital computer. Among these activities proponents of this view might include varieties of theorem proving, game playing, a limited but still extensive use of natural language, and so on. However, there is no clear characterization of those activities that are successfully mechanizable as against those that are not.

The mind is a formal system, and its organization can be understood in sufficient detail in the not-too-distant future so that machines can be built whose ability to "think" equals—if not surpasses—that of the human mind.

The mind is a formal system, but this fact says nothing about the prospects for building intelligent machines. Knowledge of the organization of the human mind of the quantity and quality necessary to build indisputably intelligent machines is just not available, nor is it "just around the corner."

Watanabe notes that one variant of the position that mental processes cannot be mechanized is the statement that digital computer simulation of the mind is doomed because analog processes are involved in the operation of the brain. As a result, it is maintained that some type of wet engineering may turn out to be inevitable. He suggests that the problem of imitating the human mind on a digital computer may not turn out to depend on inadequacies of hardware so much as on considerations of complexity itself. He then examines Simon's notions of complexity, and concludes that although Simon's decomposition of complexity into simplicity is appealing, nevertheless a task must always remain as complex as the simplest recipe available to accomplish it. The Pavlovian conditioned reflex is a simple model, but that does not mean that the entirety of a dog's behavior can in theory be explained by that model and so be mechanized—any more than the simplicity of Newton's laws implies anything about the simplicity of working out the dynamics of arbitrary physical systems. (Watanabe leaves open the question of whether or not the mind can be, in principle, mechanized.)

In the same workshop report, Miller (1974) remarks:

The various arguments that the human mind can accomplish things that a universal Turing machine cannot are intellectually fascinating, but as a practicing psychologist I have always been impressed by the finiteness and unreliability of the human mind; I think it might be easier to show that a universal Turing machine is far more powerful than a human mind—at least by the measures of power that are ordinarily used in these discussions. In any case, minds and brains are very different from Turing machines and computers, and the suggestion that either should properly include the other seems sufficiently absurd to justify turning to other matters. A more interesting question is why computers have had so little practical success in taking over information processing tasks traditionally performed by people.

Watanabe returns to say that:

It is often stated that man is a purposive agent, but this statement is immediately countered by the claim that a robot can also have a goal. Then it is contended that a human simultaneously has several competing goals; but, it would not be impossible to manufacture a robot with several concurrent goals with a random-decision making mechanism built in it. Again it is said that the goal in a robot is fixed, while the goals in a man are flexible and changing; it is however doubtful that this could be a crucial distinction. . . . The most important fact about the purposiveness of human behavior is that the multiple flexible goals in man are ultimately unified and organized into an internal value structure in accordance with maintenance and expansion of life. There is unity in multiplicity and continuity in variation. The goals in man are not just put together arbitrarily by an outside agent; they are structured as a united whole to keep and expand life. . . . The robot has goals but the man has values. . . . The value we are talking about here is not something objectivized and universal. It is, so to speak, the motive force that activates each individual.

He then goes on to argue that humans think in terms of paradigmatic symbols that need no additional rules of interpretation, whereas "machine thinking" is done in terms of abstract symbols that require an ad hoc rule of interpretation to have any relation with actual behavioral operations. Paradigmatic symbols have their places in the life-oriented internal value structure, while abstract symbols have no relation with value. It is the elicitory capability and value orientedness of paradigms that make inductive inference work. Abstract symbols, having neither of these two properties, cannot make inductive inference work without human help.

These comments over a decade ago seem still pertinent. But somehow, one feels, the machine-mind question is not well posed. What one wants out of the attempts to pose the question is clues as to possibly new technological styles of machine problem solving and designs of new machines on which to run the operations.

Limits to Knowing Brains by Simulation

The opinion that what we know about global brain operations we shall learn largely through digital simulation of brain behaviors brings one right up against the limits to simulation (Sugarman and Wallich 1983). These limits already show up in simulations of turbulence, economics, weather forecasting, airfoil eddies, and in the search for pockets of stability in the generally unstable evolution of a rapidly heated plasma. (The situation is partially captured by the joke that, in principle, we could compute exactly tomorrow's weather—but it would take a month to calculate. But the problem is worse: running current models on more and more powerful computers would, as Jared Enzler remarked, "Simply produce erroneous results faster." One hopes that more powerful machines might make possible the running of qualitatively different models that better emulate the real world.)

The Dimensionality and Speed of Brains

Anatomically at the macroscopic level, most neural networks are spatially 2-dimensional in the same sense that modern silicon chips are: the third dimension is used largely for connectivity, rather than for computation per se. The high anatomic dimensionality of the computing elements in the nervous system lies at the fan out at synaptic connections and at the molecular level where receptors "recognize" ligands as transmitters. However, to the extent that the gross hemispheric lateralizations of the human (or even mammalian?) brain represent a broken symmetry essential to its function, I acknowledge that the macroscopic 3-dimensionality is perhaps more than incidental. That is not certain; perhaps cooperativity between two large 2-dimensional arrays is what is involved, with the arrays folded in a third dimension only for efficient packaging inside a bony box that had to pass through a female pelvis at birth. Other than that, the use of the third spatial dimension may not be fundamental.

Assuming that there is some relationship between what we could mean by the dimensionality of brains and the connectivity or fan out in neuronal networks, we get a sense of the extraordinary when we consider that there are something like 10^{10} large neurons in the human brain and 10^{11} small neurons, at least as many glial cells, and that many neurons have perhaps 10^4 inputs.

Neural networks use averaging over both space and time, among other tricks, to achieve fast results from slow elements (i.e., slow compared to those in current computers). Many of those same results could be computed only very slowly by supercomputers today. Is there a hint for the designers of VLSIs in brain architectures that connect many slow elements to get fast results?

Logical Depth and Hierarchical Processes at Synapses

My reference level for a canonical neural element is the "standard" synapse, a junction between parts of two neurons with a cleft about 200–300 Å or more wide, through which one neuron, when electrically depolarized, sends a chemical transmitter to signal the others. The signal may be excitatory or inhibitory. The signalling is unidirectional, but the downstream receiver neuron—at a second, grosser level—may have an anatomical projection back onto the sending neuron, so that at this new synapse, the roles of sender and receiver are reversed, and a local loop is formed. The unit process of synaptic function at these two levels includes a third, lower level with: depolarization; transmitter synthesis, storage, release, reuptake, inactivation; second messenger release; ion channel conductance changes, etc.

At a fourth level of synaptic function, there is dendritic sprouting. A neuron under repeated activation may arborize its input structure. This is a cellular growth process, dependent upon protein synthesis and genetic constraints.

At still another (fifth) level of synaptic function, the rates of all the earlier-mentioned processes may be modulated (not just switched) by variations in the chemical field bathing the structures. This field is varied by hormones, drugs, metabolites, or nonsynaptic broadcasting of transmitters from neighboring neurons.

Thus, we have, at a canonical synapse, the possibilities of excitation or inhibition on a discrete or continuous basis, at multiple levels of structure and function. One is then curious to know: Are neuronal "logical" operations hierarchically cooperative so that neuronal junctions can be thought of as being computationally "deep" (n-dimensional, where n is at least 5)? If so, does this characteristic expose a fundamental difference between transistor (one logic-structural level) and neuronal computational networks? Is this the trick of chemical computing? (I think so, but cannot yet make the argument rigorous.)

Dynamics Versus Information in Living Systems

At least some of what we tend to think of as "informational" in biological systems has very strong dynamic aspects. During animal development, as daughter cells of the same parent cell differentiate, we see different kinds of cells arising from the same genetic message. In that process, the genome participates as an interacting component in a large dynamic network involving repressors and activators and fluctuations in chemical fields. The genome is not just sitting as a static linear array bearing a code. By the time chromosomal operations take place, the mix of information and dynamics is overwhelming. Herein lies, I think, the real question about biological computing as opposed to algorithmic, heuristic, or formal computing. The perfect embodiment of formal computing is a machine that is "transparent" to the user in the sense that its dynamics (lags, delays, switching transients, and mechanical-electrical failures) never reach his attention. The dynamics are suppressed. In great contrast, the genotype-phenotype relationship not only expresses dynamics, but bends the dynamics back toward the genome. And genes sometimes show a sort of cancerlike unrestrained replication of themselves, within the genome, making a mess out of the "linear array" image of genetic information—an image long since clouded over by the extraordinary facts of split, jumping, and nested genes.

Stuart Kauffman has modeled the genotype-to-phenotype mappings involved in cell differentiation, and this model can be extended by introduction of a fitness parameter, based on dynamics, that is a global property of the whole genome. (The physical shape of animals depends on many genes.) Brains and genes both look "wetter" than they used to.

A fundamental idea that underlies computation is that of transformations—and solid state physics offers many possibilities. But so do chemo-hydrodynamic fields. What we need, I think, is extension of the physics for soft-coupled, chemo-hydrodynamic fields in which the interactions among elements are not mass or momentum dominated; the dynamics are governed by kinematic interactions. Biological "information" does not arise directly from kinetics, nor does it consist only of symbol strings. It arises from kinematic interactions—length and time (but not mass) dependent. That is the first-level "abstraction" of a chemo-hydrodynamic field, and I am suggesting that it is sufficient for creating a nonkinetic class of interactions that might justifiably be called "communicational." The information is in singularities that arise at this level.

Digital computers suppress kinetics, and emphasize symbolic string processes. Classical Newtonian systems interact kinetically (spatial coordinates, mass, length, and time all participating). Thus, the digital computer represents a nearly pure information system; Newtonian mechanical systems represent purely dynamic (kinetic) systems. Biological systems lie in between, governing their dynamics through interacting kinematic flow field singularities, and through structures acting as constraints.

Hybrid Systems: Man-Machine

Gel'fand and Tsetlin (1962) approached the problem of control of complex systems through the mathematical study of animal locomotion. They ended up having to consider the problem of finding the minimum of a function of many variables. In a footnote, they comment that Ulam proposed to deal with such problems by constructing a hybrid between man and machine. The machine shows on a screen some 2-dimensional section of the function in question, and the man—looking at this relief—decides how to proceed further (i.e., what sections to take, what to look at under magnification). Here, the human brain is used to parse the problem into elements that the machine can compute rapidly, whereas direct application of algorithmic methods to the problem in the first instance would not be feasible in practice. Gel'fand and Tsetlin see the human as parsing the problem according to a nonlocal search method (that they formalized as the "ravine" method) that allows one to use deeper properties of the organization of the function than merely its local behavior. The approach applies to what they call "well-organized" functions.

A well-organized function is one that permits a division of its parameters into two groups, the first consisting of parameters whose change leads to significant changes in the "evaluation function" of the mathematical system. (It is the evaluation function that is to be extremized, and this function is dependent on time.) The second group of parameters contains a small group of variables such that a change in the variables of this group leads to a relatively small change in the evaluation function. If the variables of the first and second groups are equally important, then the function is not well organized. The approach to well-organized functions by the nonlocal ravine method was compared with those through local methods such as gradient or steepest descent, and it was shown that the nonlocal approach deals much better with functions of many variables—something that we may suppose brains do well when, for example, they coordinate walking or assure successful gait transitions through potential instabilities to stable jogging or running.

How brains parse flow fields is as important to know as are their (mysterious) data structures! Both presumably hold the clues as to what brains do so well that Turing machines don't.

Fifth-Generation Machines

No fundamental physical obstacles are likely to impede the development of new "fifth-generation" digital computers that are larger in computational capacity, smaller in spatial dimensions, and faster (with overall cycle times of less than four nanoseconds) than our current best machines (Moto-Oka 1982). However, it remains to be proven that there can be algorithms for scheduling highly parallel machines that can guarantee the running of a complex program. If there are about a thousand or more separate processors being scheduled, as is planned for the fifth generation, then execution graph analysis (displaying the number of steps each processor can carry out before it must wait for the answer from another one) shows that it can be highly likely that many or even all of the processors will ultimately end up waiting—leaving as the only process ongoing in the machine futile attempts by the master control program to make something further happen. This locking up of the computation is not highly architecture dependent, but arises from the structure of the problem itself. Thus, there may be too much optimism about the probable early success of the fifth generation of computers, and not enough critical review of how we will program these machines.

Execution-Driven Versus Program-Driven Processes

Current plans for advanced digital computers still emphasize program-driven computations. It is noteworthy that the evolution of dynamical and informational processes in animal development is execution driven and not program driven. The information in the genome specifies linear arrays of amino acids in proteins, and the genetic code constitutes the simplest "machine language" of the biological system. However, that coding is not sufficient to generate an organism. During development, each stage is initiated by the structures completed at the preceding stage that were not present in advance as an option. If the preceding stage is not executed, the subsequent stage does not occur correctly or at all.

I suggest that in the style of development of a biological organism, we can find the proper style for advanced serial and parallel processing and its scheduling. In biological systems, the execution not only sets the stage for the next round of development, but also feeds back to the genome to release a new round of programs. The "start" and "stop" timing codes are not clock driven nor driven by other parts of the program; they are driven by the dynamical accomplishments following preceding stages of the program. Therein may lie the deepest clue for advanced computation. The designers of digital computers may soon have to look again to biology for inspiration.

The reader will see that I have no hope for or interest in explaining the adaptive characteristics of living systems through use of terms or concepts borrowed from artificial intelligence or computer science. Indeed, I think these latter two endeavors do better by turning their attention to biology. Not only do many living systems perform splendidly without nervous systems, but even those that have nervous systems (and other "information-rich" subsystems such as immune systems, genomes, hormone receptors) present as their dominant microscopic processes the making, breaking, and exchange of chemical bonds. Thus, whatever we choose to anthropomorphize as "information" in them is an emergent property from their dynamics. The question then is: Do we have to pay attention to the dynamics to get at the computationlike, evolutionary, and adaptive properties? Certainly, it is true that modern digital computers are designed so that their switching dynamics are suppressed and not noticed by the user of the machine. For the user, the machine is seemingly only a follower of (arbitrary) rules regarding handling and transformations of symbol strings. Is it the case that in explaining biological system dynamics to ourselves, we can similarly ignore dynamics? I claim that the answer is: No.

The tension between an informational (or algorithmic or computer science or finite-state automaton) view of computing processes and the dynamical view was graphically caught in the cover photograph on Volume 27, Number 3 of the *Norelco Reporter,* December 1980 (Phillips Electronic Instruments, Inc., 85 McKee Drive, Mahwah, New Jersey 07430). The photograph shows a scanning electron micrograph of an ant carrying in its jaws a silicon chip. Here, the world of carbon and the world of silicon intersect dramatically, and the similarities and differences are striking.

The chip is 64 K; the ant has 64,000 (approximately) "neurons" (these cells are not like most mammalian neurons, but nevertheless are elements of the nervous system). Both the chip and the ant are biological objects— the ant arising out of biological evolution, and the chip arising as an artifact or byproduct of biological operations. Both are intentional systems (though in different senses). There the similarities cease, and the differences become conspicuous.

The ant is self organized, whereas the chip was carefully fabricated from a photolithograph. The ant is not provably a state-determined system. The chip circuit is single-level connected, and the ant has hierarchical connections (what I have referred to as structural depth associated with logical depth). The chip has speed but limited connectivity; the elements of the ant nervous system are slow, but have extensive fan out. In the ant, the past is prologue in a historical, biological evolutionary sense—not just the present state sense. Finally, the ant has a microstructure unlike its macrostructure (it is not a crystal) specialized for storing aspects of its tribal past. The ant works by making, breaking, and exchanging chemical bonds. In the ant, the components are not independent of the problem being solved, but in the chip, they are.

In conclusion, the operations of the ant and the chip could hardly be more different. We lose insight if we use a solid-state, silicon-based, computer metaphor for the ant. Instead of switches we find, in biology, cooperating and competitive dynamic processes, dynamic regulation with circular causality instead of negative feedback with references, set points, and comparators; we have epigenetic effects as powerful as genetic effects; we find that random events not only may affect immediate operations (for example, by driving the system through a bifurcation), but that these consequences are accumulated to bias future possibilities. In that sense, chance becomes creative necessity (Monod 1971).

Two Exhibits: How Plants Find Up and Down, and How Brains Process Odors

Gravitropic Responses in Plants: Clever Dynamics

When plants grow on the surface of the earth, roots go down and stems go up. There is no nervous system to solve that problem. Admittedly, some systems, such as underground rhizomes, actually grow horizontally, while others, like the branches of a tree, grow at a fixed angle to the vertical. The prop roots of corn grow downward at a 45-degree angle from the vertical, the branch roots usually differ in their orientation from the main roots, and these behaviors need not be fixed.

An 80-foot-tall cedar tree near my mountain cabin has a triple top (the original top having been killed by lightning), and four large main branches that leave the trunk horizontally, extend away from it about two feet, and then acutely take a right angle turn to the vertical and grow straight up, giving the whole tree an appearance rather like that of a menorah. Other cedars nearby have the normal habitus—all main branches start out horizontal and remain so. Question: In the case of the unique tree, how did the branches know about the injury to the central growing tip at the top, and try to become tops themselves by managing to turn right angles simultaneously (if, as is likely, length is proportional to time in the tree)? For that matter, how does any plant know up from down to any degree?

According to updated versions of the now classic Went-Cholodny theory, the gravitropic signal, when a vertical stem is tipped to horizontal, is an asymmetric distribution of the growth hormone, auxin. Certain cells in the plant, the statocytes, contain dense, starch-containing plasmids or amyloplasts that sediment within several minutes when the orientation of the plant organ is changed with respect to the vertical. Amyloplast displacement appears to be the first step in the "perception" of the earth's gravitational field. The plasmids have been called statoliths. The displacement of the amyloplasts causes contact with other organelles within the statocyte, such as the endoplasmic reticulum, the cytoskeletal components, or the plasmalemma. An unknown process triggered by that contact is probably the first step in the transduction of the gravitational stimulus into a signal that is transmitted to the growing part of the organ. Then, the upper surface has its growth slowed or inhibited, and the lower surface has its growth rate accelerated, and an upward bend occurs if the stem has been held horizontal when it would normally have gone straight up. The whole thing is presumably done with only a few chemical signals including indoleacetic acid, protons, calcium ions, calmodulins (maybe), and perhaps, cAMP and an electrical potential difference. The mechanism is not understood, and there are many controversies (e.g., there are starch-free mutants with gravitropism), but it is chemical kinetics the whole way. The same can be said for thigmotropic and phototropic responses of plants. In every case, organized changes in structure and function are brought about by physical variables as stimuli (gravity, touch, wind motion, light, temperature, and injury). Not even Joyce Kilmer ("I think that I shall never see a poem as lovely as a tree") would ascribe an internal representation of the environment to the guts of the plant. And to make the situation even more startling, not only do plants lack nervous systems, but they don't even have (yet discovered) receptors for their few hormones!

What is needed to comprehend plant physiology is a dynamical view of carbon chemistry that emphasizes the making, breaking, and exchange of bonds at relatively low ev energies, under restricted ranges of temperature, with emphasis on six main processes: activation or inhibition, cooperation or competition, chemical complementarities, and broken symmetries. Those are the same processes that support perception and response to stimuli in animal nervous systems, as will be described.

How Brains Recognize Smell

During a 12-year exhaustive search for the electrical patterns in multiple recording channels (up to 64 channels per olfactory bulb) chronically implanted in the brains of conscious animals, using multiple statistical pattern-extraction methods including temporal and spatial spectra, Freeman and Skarda (1985) made the following conclusions about how animals might recognize smells:

1. The typical pattern of 64-channel bulbar recordings in aroused animals was common to all channels, and each animal had its characteristic spatial pattern of phase and amplitude that, like signatures, were easily recognized but never twice identical.
2. The details of the shapes of active foci varied erratically and unpredictably, as did the mean rms amplitude from each burst to the next.
3. When up to 24 odorants were presented to each animal, there were no consistent differences either in phase or in amplitude in these recordings that depended on the odorants.

4. When animals were conditioned according to several paradigms, using odor as the conditioned stimulus, the control and odor bursts of electrical activity did not differ significantly, even though the behavior of the animal did differ.

5. It was discovered that the matrix of amplitudes of the dominant oscillatory component of the electrical signals carried odor-specific information, and this information was present in every channel. (This discovery terminated 12 years of unsuccessful attempts to correlate features of activity on particular electrodes—spatial coordinates—with the presence of particular odorants.)

6. The spatial extent of coherence far exceeded the spatial range that could be accounted for by volume conduction alone.

7. The bulbar mechanism was inherently nonlinear.

8. A dynamic model of the bulb, embodying 64 coupled sets of differential equations, could incorporate the static nonlinearity of the system. The several coefficients in the model represented time, space, and gain parameters, and these could be adjusted to fit the physiological data, such as that summarized in item 9.

9. The odor-specific information in the recordings that served to classify bursts correctly with respect to stimulus condition was uniformly distributed among the amplitude coefficients of the dominant component of coherent, higher frequency bursts. In the presence of a certain odor complex, the dynamics underwent a bifurcation from a chaotic mode to a mode resembling (a generalized) limit-cycle performance. The details of phase, center frequency, and temporal amplitude and frequency modulation were insignificant. The existence of each stable pattern depended upon a successful learning process that resulted in emergence of discriminatory behavior, and for each discriminated odorant, a unique limit cycle attractor formed, which was distinguished from other members of its class by its basin. Changes could be imitated by a decrease in phase, frequency, and decay rate of the dominant damped cosine fitted to the responses in the model. Simulation of this pattern, with the solutions to piece-wise linear and nonlinear differential equations modeling the dynamics of the bulb (according to a Hebb nerve cell assembly), showed that a small increase in the coefficients representing the coupling strength of "synapses" from excitatory neurons to other excitatory neurons was sufficient to account for the results. A modest increase of 40% in the coupling strength increased the sensitivity of local oscillators 40,000-fold. In the model (and, by inference, in the bulb), the stereotypic output pattern was global in the sense that it involved all elements, but each local region, such as the fraction

of the bulb covered by an electrode array, contained all of the information, at reduced resolution compared with the whole. In that respect, though not in others, bulbar output resembled a holographic storage pattern.

From these remarkable details consisting of experimental results and computer simulations, Freeman and Skarda concluded that synthesis in the olfactory bulb is through bifurcation from a dynamic chaotic state of sensory input to an oscillatory state of output to the olfactory cortex. From this, animal behavior was governed. They believed that the "purpose" that we might impugn to such animals does not emerge in the form of a representation of a future state, but consists of the existence of a hierarchy of topological attractors in the dynamics of large masses of neurons, such that the behavior of the central nervous system—and therefore, the animal—evolves from the present state with a tendency to arrive at a definite future state. No representation is involved; the remote origin of purpose of behavior lies in the embryology of the brain, wherein form and function interact to generate a predictable sequence of changes such that goals are implicit. Purpose in this sense is best conceived in mathematical terms as a trajectory of directed movement in phase space through an ordered sequence of basins and bifurcations.

Is Evolution a Game?

I have attempted to challenge the all-too-easy, conventional wisdom that the key to understanding evolutionary problem-solving processes in biology can be found in algorithms and information science metaphors. An extreme case of this (wrong, I think) position can be found in Poundstone's engaging book, *The Recursive Universe* (1985). It presents the evolution of the universe and of self-reproducing life patterns in the context of the Game of Life, introduced by John Horton Conway via Martin Gardner's columns in *Scientific American* (October 1970 and February 1971). As Poundstone says:

[This is] a fantastic computer game called Life. It is played by collegiate computer hackers, by distracted employees with large computers, by filmmakers experimenting with computer animation, and by sundry others on home computers. . . . The Life screen, or plane, is a world unto itself. It has its own objects, phenomena, and physical laws. . . . Conway further showed that the Life universe—meaning by that a hypothetical infinite Life screen—is not fundamentally less rich than our own. All the variety, complexity, and paradox of our world can be compressed into the two dimensions of the Life plane. There are Life objects that model every precisely definable aspect of the real world. (Note that Abbott first introduced such ideas in 1884, and Dewdney has elaborated them in 1984.—F.E.Y.)

Life's rules are marvelously simple. . . . The rules determine everything so that the game plays itself. Each cell may be in one of two states. The states are called on and off, 1 and 0, occupied and empty, or live and dead. . . . At any instant the Life universe can be described completely by saying which cells are on. . . .

Poundstone goes on to insist that a living system meet the requirements of von Neumann's theory of self-reproducing automata (1966):

1. A living system encapsulates a complete description of itself.
2. It avoids the paradox seemingly inherent in item 1 by not trying to include a description of the description in the description.
3. Instead, the description serves a dual role. It is a coded description of the rest of the system. At the same time, it is a sort of working model (which need not be decoded) of itself.
4. Part of the system, a supervisory unit, "knows" about the dual role of the description, and makes sure that the description is interpreted both ways during reproduction.
5. Another part of the system, a universal constructor, can build any of a large class of objects—including the living system itself—provided that it is given the proper directions.
6. Reproduction occurs when the supervisory unit instructs the universal constructor to build a new copy of the system, including a description.

It is true that a purely behavioral (dynamic) definition of life is handicapped by the fact that some forms of life have little behavior. Some bacteria and spores do not exhibit irritability, or even metabolism for extended periods. Poundstone argues that the cogency of the information-theory definition of life is better seen when applied to problematic cases: "Whenever biologists try to formulate definitions of life, they are troubled by the following: a virus; a growing crystal; Penrose's tiles; a mule; a dead body of something that was indisputably alive; an extraterrestrial creature whose biochemistry is not based on carbon; an intelligent computer or robot." To his list, I would add flames and the tardigrade that is unmistakably alive when wet and dead when dry, and can cycle back and forth between the two states (Yates 1986).

In other discussions, I have emphasized the difficulty of relating information processes and dynamics in living systems (Yates 1982a,b;

Yates and Kugler 1984; Yates 1985; Yates 1986). Pattee (1977, 1982) has consistently called attention to this problem. In examining the use of concepts such as "symbol," "referent," and "meaning" in the descriptions of biochemical and evolutionary phenomena, he sees cells as symbol-matter systems transforming strings from a small set of elements, by rules, into other strings, in turn transformed by laws into functional machines. In his account, information and dynamics are complementary.

The information metaphor that I claim is inimical to clear thinking in attempts to reduce biology to physics and to comprehend the evolutionary aspects of biological adaptations as "computations" will probably never disappear; it is too convenient a shorthand. But there is little or no more substance in the statement that "genetic programs direct development" than there is in the statement "balloons rise by levity" (Yates 1986). The unsatisfactory state of the evolutionary biology of a single organism (that is, of developmental biology) and its distance from natural science is perhaps being relieved by further applications of topological approaches to the problems of morphogenesis. Thom (1986) has commented that "All modern biological thought has been trapped in the fallacious homonym associated with the phrase 'genetic code' and this abuse of language has resulted in a state of conceptual sterility, out from which there is little hope of escape." Fortunately, he offers some escape through a new extension of his original account of 1972.

Working with four morphologies of flow (birth, stopping, confluence, or ramification), Thom shows that the ancient millwheel expresses many dynamic features needed to describe embryonic morphogenesis in metazoa: canalization of dynamics, coupling to a potential to obtain a direct flow, entrainment, nonlinear catastrophic escapements during a (fictive) retroflux, dissipation of free energy, and periodic behavior. He argues that life is itself a spatially and biochemically canalized phenomenon with dynamics shaped by membranes, cytoskeletons, and macromolecules. Cell replication requires duplication of a singularity in a flow field undergoing continuous deformation. (Note the similarity to the Gibsonian view of specificational information.) In biology, evolution is irreversible. All forms are the descendants of what has been, not of what might have been; all that will be is the descendant of what is—evolution is historical rather than predictive. In living systems, the laws of physics are constrained by structures and processes that themselves are the result of evolution, though the laws remain unchanged. The constraints are multilevel, and press upon biological processes from above and from below. In those constraints are the history and the operation of the second law of thermodynamics, which, in effect, says, "Constraints shall not last." The progressive loss of constraints (in open as well as closed systems) is the process of aging.

Summary

We find today an enrichment of computer science by borrowings from biology, but not the reverse. In computer science, the purpose of the so-called "genetic algorithms" is to choose a scheduling for a time-sharing system, without the usual tuning or interventions necessary during unusual demands. Adaptive solutions modify the underlying algorithms dynamically in response to the particular demand characteristics of the problem at hand. For example, in the famous traveling salesman problem (finding the shortest tour of n cities in order to minimize traveling expenses for a salesman), genetic algorithms attempt to evolve the shortest tour (the optimal solution) by encoding each tour as a finite automaton, killing off poorly behaving automata (long tours) and cross-breeding and mutating the survivors.

The need for programs that adapt or fine tune themselves and grow in interesting ways is clear. Here, computer scientists have looked to biology for analogies on how to create adaptive systems. But technological "genetic" algorithms are limited in a number of ways. In several cases, including the traveling salesman problem, the difficulty has been in finding appropriate representations that can use the approach. In other cases, the algorithms have not performed better than random search.

In great contrast to the genetic algorithms of computer science, we find that living systems evolve the means to solve problems rather than evolve algorithms directly. Animals satisfice rather than optimize. In biological systems, surface complexity does not necessarily arise out of deep simplicity, whereas in technological systems, and in the physical universe more generally, this is thought (or hoped) to be so (Weinberg 1985). Biological systems, whether individual organisms or species, take advantage of side-effects, emergent properties, and unexpected interactions in such a way as to support the development of new lines of subsystems. With respect to the evolution of species, S.J. Gould has coined the term "exaption" to describe the phenomenon that a structure associated with one function later becomes part of a new function. The first fishes did not have jaws, but those bones were already present in ancestors, doing another job—they supported a gill arch just behind the mouth. They were well suited for their respiratory role there; they had been selected for this alone. Nevertheless, under changed circumstances, they proved admirably "preadapted" to become jaws. Similarly, in some insects, structures that appeared first to accomplish heat exchange later functioned as wings for flight, and in nervous systems, we see many different ways in which neurons can form relations with other neurons; the possibility for unexpected side-effects and emergent properties has been great. Relationships are not fixed or immutable. The hardware changes as the problem changes.

The trick of life is the breaking of structural and dynamic symmetries. (Anderson (1972) has given a clear account of the power of broken symmetry in constructing a diverse universe.) Wherever we look into changing behavior of organisms (the geotropic plant changing the direction of growth of a branch; the self-organization of a dynamic pattern distributed across a neuronal field as the manifestation of detection of a particular smell; the evolution of jaws from gill arches in fish), we see that change in structure is change in function, and change in function sustains, maintains, or changes structure. Out of that intimate joining of particular (not general purpose) structure and function, we find the adaptive solutions to problems posed as the system evolves. Structure and function are revealed as information and dynamics melded in the context of evolutionary processes that are problem solving. 0's and 1's do not occupy center stage. They are a dropping of our culture.

Bibliography

Abbott, E.A., 1884: *Flatland: A Romance of Many Dimensions*, Signet Classic (1984), New American Library, New York.

Anderson, P.W., 1972: More is different, *Science* 177:393–396.

Carello, C., M.T. Turvey, P.N. Kugler, and R. Shaw, 1984: Inadequacies of the computer metaphor, in M.S. Gazzaniga (ed.), *Handbook of Cognitive Neuroscience*, Plenum Press, New York, 229–248.

Churchland, P.M., 1982: Is thinker a natural kind? *Dialogue* 21:223–238.

Conrad, M., 1974: Evolutionary learning circuits, *Journal of Theoretical Biology* 46:167–188.

Conrad, M., 1983: *Adaptability—The Significance of Variability from Molecule to Ecosystem*, Plenum Press, New York.

Dewdney, A.K., 1984: *The Planiverse—Computer Contact with a Two-Dimensional World*, Poseidon Press, New York.

Freeman, W.J., and C.A. Skarda, 1985: Spatial EEG patterns, non-linear dynamics and perception—the neo-Sherringtonian view, *Brain Research Review* 10:147–175.

Gel'fand, I.M., and M.L. Tsetlin, 1962: Some methods of control for complex systems, *Russian Mathematical Survey* 17:95–116.

Hinton, G.F., T.J. Sejnowski, and D.H. Ackley, 1984: *Boltzmann machines—constraint satisfaction networks that learn*, Carnegie-Mellon University, Pittsburgh, Pa.

Hopfield, J.J., 1982: Neural networks and physical systems with emergent collective properties, *Proceedings of the National Academy of Sciences USA* 79:2554–2558.

Kirkpatrick, S., C.D. Gelatt, Jr., and M.P. Vecchi, 1983: Optimization by simulated annealing, *Science* 220:671–680.

Miller, G.A., 1974: Proceedings of the workshop on possibilities and limitations of artificial intelligence, *IEEE Transactions, Systems, Man and Cybernetics* SMC-4:88–103.

Monod, J., 1971: *Chance and Necessity*, A.A. Knopf, New York.

Moto-Oka, T. (ed.), 1982: *Fifth Generation Computer Systems*, North Holland Publishing Co., New York.

Pattee, H.H., 1977: Dynamic and linguistic modes of complex systems, *International Journal of General Systems* 3:259–266.

Pattee, H.H., 1982: Cell psychology—an evolutionary approach to the symbol-matter problem, *Cognition and Brain Theory* 5:325–341.

Poundstone, W., 1985: *The Recursive Universe—Cosmic Complexity and the Limits of Scientific Knowledge*, William Morrow and Company, New York.

Sugarman, R., and P. Wallich, 1983: The limits to simulation, *IEEE Spectrum*, April, 36–41.

Thom, R., 1972: *Stabilite Structurelle et Morphogenese—Essai d'une Theorie Generale des Modeles*, Benjamin, Reading, Mass., and Intereditions, Paris, France.

Thom, R., 1986: Organs and tools—a common theory of morphogenesis, in J. Casti (ed.), *Complexity, Language and Life—Mathematical Approaches*, IASA, Springer, Luxembourg.

Turvey, M.T., and P.N. Kugler, 1984: A comment on equating information with symbol strings, *American Journal of Physiology: Regional, International Compilations of Physiology*, R925–927.

von Neumann, J. (with A.W. Burks), 1966: *Theory of Self-Reproducing Automata*, University of Illinois Press, Urbana, Ill.

Waner, S., and H.M. Hastings, 1987: Evolutionary learning of complex modes of information processing (Chapter 7, this volume).

Watanabe, S., 1974: Proceedings of the workshop on possibilities and limitations of artificial intelligence, *IEEE Transactions on Systems, Man, and Cybernetics*, SMC-4:88–103.

Weinberg, S., 1985: Origins, *Science* 230:15–18.

Yates, F.E., 1982a: Systems analysis of hormone action—principles and strategies, in R.F. Goldberger and K.R. Yamamoto (eds.), *Biological Regulation and Development, Vol. 3A—Hormone Action*, Plenum Press, New York, 25–97.

Yates, F.E., 1982b: Outline of a physical theory of physiological systems, *Canadian Journal of Physiology and Pharmacology* 60:217–248.

Yates, F.E., 1984: *Report on Conference on Chemically-Based Computer Designs*, Crump Institute for Medical Engineering Report CIME TR/84/1, University of California, Los Angeles, Calif.

Yates, F.E., 1985: Semiotics as bridge between information (biology) and dynamics (physics), *Recherches Semiotiques/Semiotic Inquiry* 5:347–360.

Yates, F.E., 1986: Quantumstuff and biostuff—a view of patterns of convergence in contemporary science, epilogue to *Self-Organizing Systems—The Emergence of Order,* Plenum Press, New York.

Yates, F.E., and P.N. Kugler, 1984: Signs, singularities and significance—a physical model for semiotics, *Semiotica* 52:49–77.

4. Selective Networks and Recognition Automata

Introduction

One of man's most ancient and perplexing problems has been to understand the nature of his own mental processes. The invention of the computer has, for many, suggested a promising approach based on the study of information processing and artificial intelligence. However, we would like to suggest that animals, including ourselves, face some problems that computers do not have to face: we are exposed to an unpredictable environment and must respond adaptively in order to survive. We can get information only in real time through the senses, not precoded on a magnetic tape. Above all, we must function without a pre-established program that would specify appropriate responses for all contingencies. In other words, we have to decide for ourselves what the problem is; it is not given to us.

In this chapter, we shall consider an essentially biological approach to the problem that draws its inspiration from the Darwinian theory of natural selection and that stands in marked contrast to the information-processing approach. As we shall see, selective systems provide a way to overcome the principal difficulties of the information-processing approach: the needs to provide pre-established external information in the system along with a code for representing that information and algorithms for processing it. These needs arise from the formal separation of hardware and software in information-processing theory, a distinction that has no place in a biological theory of the mind. We believe that category formation and categorization lie at the heart of the matter; that memory and other higher brain functions

Reprinted with permission from *Annals of the New York Academy of Sciences,* Volume 426, *Computer Culture: The Scientific, Intellectual, and Social Impact of the Computer,* 1984, pp. 181–201.

depend on associations between stimuli that are intimately related to category formation; and that the way we perceive categories involves combinations of functions that are quite unlike those used to date for computer category perception. Accordingly, this chapter will concentrate on problems of pattern recognition and category formation. We will review a theory for biological pattern recognition that is based on a selective principle, will describe briefly a pattern-recognizing automaton that we have constructed based on this principle (Edelman and Reeke 1982) and will try to relate the results obtained with this automaton to the broader questions we have raised.

We begin with the seemingly spontaneous perception of pattern or Gestalt that occurs whenever we recognize objects in the visual world. How we are able to do this is a deep problem, as suggested by two simple illustrations.

In Figure 4.1, we reproduce a fragment from M.C. Escher's *Metamorphosis II*. The tower in the center is perceived as a rook in a chess game when viewed in the context to its right, but as part of an Italian town when viewed in the context to its left. The illustration is a kind of visual joke, deliberately ambiguous, but it does illustrate the importance of context in shaping our perceptions.

Figure 4.2 gives another example, again somewhat extreme. It illustrates the concept of a polymorphous set, a set with no singly necessary and jointly sufficient conditions for set membership. Everyday concepts often have polymorphous characteristics—they are not as logical as we would like to think. For example, the reader might attempt to construct a list of necessary and sufficient conditions to define an object as a piece of furniture. Quite likely, no such conditions can be found—only a list of characteristics, some of which are shared by any particular piece of furniture. Or, consider the everyday definition of a bird. Most people will mention that "a bird flies," even while knowing perfectly well that there are birds that do not fly. A large body of psychological literature (for a review, see Smith and Medin 1981) suggests that we identify categories by a mixture of two procedures: comparison with exemplars, and probabilistic feature matching. Not only that, but we slide back and forth between these two methods, depending on the situation at hand. One of the main points of this chapter is that both methods in combination are necessary for more complex functions such as the formation of associative memories, not because we are careless or illogical, but because the need for them is built into the structure of the underlying system. For this reason, both types of categorization are present in the automaton we will describe.

It is important to understand that learning need not necessarily be a factor in recognition, although it is often useful. What is required is that a response be associated with a pattern of sensory information corresponding to an object or class of objects in the environment, and that the response be adaptive for the organism. Conventional responses, such as names, can be acquired only through learning, but before we can learn, it is necessary to have categories or, at higher levels, concepts to which the learned responses can be attached. The system we will discuss carries out this more primitive form of pattern recognition that does not involve learning. It incorporates the ability of biological systems to respond to stimuli never encountered before, yet has no program that embodies information about particular stimuli.

To see how this can be done, consider some facts about biological recognizing systems and what they imply. The brain comprises a network of interconnected neurons with apparently a high degree of parallelism in their operations. The speed and dynamic range of these neurons are limited, suggesting that the network cannot carry out algorithmic calculations involving large number of steps and high-precision arithmetic. We also know something about the number of neurons and how they are connected. Specialized regions are seen in the brain, consistent with the notion that multiple networks with different functions may need to interact to generate more complex functions. One sees highly overlapping arborizations of individual neurons, with apparently a large degree of variability, even in genetically identical individuals. Such a level of variability in wiring would be a severe nuisance in a digital computer; here, one would like to think nature has adopted it as part of the solution to the problem. Finally, no single neuron appears to be indispensable to memory (Lashley 1950), strongly suggesting that patterns of response in collections of neurons, not individual responses, are what is important.

All of these observations—particularly the necessary presence of variation in the structure of the system and its need to respond adaptively to an unpredictable environment—are the hallmarks of a selective system, the same ingredients that one sees in the working of evolution itself. We suggest that the brain is, in fact, a selective system operating in somatic time and that selection provides the exquisitely tuned adaptability needed for survival in a hostile environment.

The Neuronal Group Selection Theory

The conditions needed for a selective system are: (1) a collection of variant entities, or repertoire, capable of responding to the environment; (2) sufficient opportunities for those entities actually to encounter the environment; and (3) a mechanism to enhance or amplify differentially the numbers or strengths of those entities whose responses to the environment are in some sense adaptive. All of these elements are present in the neuronal group selection theory (see Figure 4.3) (Edelman 1978). The recognizing elements are postulated to be groups of from 50 to 10,000 interconnected neurons. These groups are formed during development, prior to exposure to any sensory stimulation, and their connections, once established, remain stable thereafter. The groups are connected in networks to form repertoires capable of encountering sensory information transmitted along the network connections. When the response of a group happens to be adaptive for the organism as a whole, the strengths (transmission efficiencies) of the connections of that group are modified in such a way that the response of the group is faster or stronger to future encounters with the same or similar stimulus patterns.

In such systems, there is necessarily a relationship between the number of recognizing groups in a repertoire and the specificities of the individual groups. If recognition is overly specific, there cannot be enough groups in a finite repertoire to recognize all possible stimuli, and the system must fail; similarly, if specificity is too broad, similar but significantly different stimuli cannot be distinguished, and, again, the system must fail. The specificities must therefore be intermediate, but this implies that several groups may respond more or less well to any given stimulus. This phenomenon, which we call degeneracy (see Figure 4.4), is critical to an understanding of selective recognition systems. Sufficient degeneracy assures that there will be some response to any conceivable stimulus, and, in fact, that more than one group will respond to any stimulus, assuring the necessary degree of functional redundancy to make the system "fail-safe" against the failure of individual groups.

Figure 4.5 illustrates how a degenerate recognizing system becomes capable of responding to any stimulus, even a novel one, once it becomes sufficiently large. The crucial assumption is that there is a small, but finite, probability that any single group will respond to any given stimulus. This assumption seems reasonable for a network of groups connected so that each of them receives at least some part of the total pattern of sensory input elicited by each stimulus. A change in the recognition specificity serves to shift the entire curve in Figure 4.5 to the left or to the right, more groups being needed if the probability of response of each individual group decreases. In practice, the probability of response, p, must be set according to the level of specificity needed to deal with a particular ecological niche, and the required size of the network, N, follows.

This theory is compatible with the biological facts we have summarized, and suggests a way that organisms can make use of the unavoidable epigenetic variability in neural nets to recognize and classify stimuli even in the absence of classical reinforced learning. Its relevance to brain function will, of course, have to be decided by experiment; already, evidence in support of the theory is beginning to accumulate (Edelman and Finkel 1984). However, independent of its biological applicability, one can, with a computer model, test the consistency of the theory and explore its power to solve real classification problems. Such a model can serve to focus experimental questions for biologists, and, at the same time, point the way to the construction of artificial pattern-recognizing systems employing the same principles. We believe that the model we have constructed, which we call "Darwin II," represents a new kind of pattern-recognizing automaton in that it can function without a program and without forced learning to recognize stimuli, classify them, and form associations between them. The construction of this automaton will be discussed in the next section.

Darwin II

In devising the Darwin II model, we were guided by a number of ground rules. It was obvious that the system should be a network. The nodes of this network are the recognizing elements of the model, corresponding to the groups of neurons postulated by the theory. We decided to model these groups at the functional level, not at the level of detailed electrophysiological properties. Each group in the model has a state corresponding to its level of activity. A group's state is dependent only on its present inputs and past history. The groups are able to transmit their state variables to other groups along the connections of the network, which are analogous to the synapses in a nervous system. As in the adult central nervous system, the connectivity of the network, once established, is not changed. However, the connections' strengths can change, and it is these changes that provide the mechanism for the amplification of response required by the theory. Finally, one of the most important rules—one that distinguishes Darwin II from systems based on "frames" or "conceptual networks"—is that there can be no specific information about particular stimulus objects built into the system. Of course, general information about the kinds of stimuli that will be significant to the system is implicit in the choice of feature-detecting elements we have made—this is akin to the choices built into organisms by their evolutionarily determined programs.

The overall plan of Darwin II is shown in Figure 4.6. At the top is an "input array" where stimuli are presented as patterns of light and dark picture elements on a lattice. (We typically use letters of the alphabet on a 16 × 16 grid, but any 2-dimensional pattern is acceptable.) The system proper is below the input array. It consists of two parallel concatenations of networks, each with several subnetworks or repertoires (indicated by boxes). These operate in parallel, and "speak" to each other to give a function not possessed by either set alone. The two sets of networks are arbitrarily named "Darwin" and "Wallace" after the two main figures in the description of natural selection. The Darwin network (left) is designed to respond uniquely to each individual stimulus pattern, and loosely corresponds to the exemplar approach to categorization. The Wallace network (right), on the other hand, is designed to respond in a similar fashion to objects belonging to a class, and loosely corresponds to the probabilistic matching approach to categorization. Darwin and Wallace have a common level structure. Each has connected to the input array a level that deals with features, and below that, an abstracting or transforming level that receives its main input from the first level. Output may be taken from these networks at the bottom.

The first part of Darwin is the **R** or "recognizer" repertoire. It has groups that respond to local features on the input array, such as line segments oriented in certain directions or with certain bends, as suggested by the inset (see Figure 4.6). Sets of these feature detectors are connected topographically to the input array so that the patterns of response in **R** spatially resemble the stimulus patterns. Connected to **R** is a transforming network called **R-of-R** ("recognizer-of-recognizers"). Groups in **R-of-R** are connected to multiple **R** groups distributed over the **R** sheet, so that each **R-of-R** group is capable of responding to an entire pattern of response in **R**. In the process, the topographic mapping of **R** is not preserved; as a result, **R-of-R** gives an abstract transformation of the original stimulus pattern. (Note that if the stimulus undergoes a change—such as a translation to a new position on the input array—the pattern of response in **R-of-R** will be quite different. It is Wallace that deals with this translation problem. **R-of-R** is concerned with *individual* properties of a stimulus, and these include its relation to the background.)

Wallace begins with a tracing mechanism designed to scan the input array, detecting object contours and tracing along them to give correlations of features that reveal the presence of objects as single entities and their continuity properties and that respond to some of their characteristics, such as junctions of various types between lines. In this respect, Wallace works something like the eye does in rapidly scanning a scene to detect the objects present. The result of the trace is that a set of "virtual" groups ($G_1 \ldots G_{27}$ in Figure 4.6) are excited according to the topology of the input pattern. These groups are called "virtual" because their input does not involve ordinary synaptic connections. The virtual groups are connected in turn to an abstracting network, R_M, which responds to patterns of activity in the trace in much the same way as **R-of-R** responds to patterns of activity in **R**. Because the trace responds to the presence of lines or junctions of lines with only little regard for their lengths and orientations, R_M is insensitive to both rigid and nonrigid transformations of the stimulus object, and tends to respond to class characteristics of whole families of related stimuli.

The **R-of-R** and R_M networks are connected together by the reciprocal cross-connections shown at the bottom center of the figure. These connections are re-entrant in that they connect one part of the system to another part of itself rather than to the outside; they provide the mechanism needed for the system to display associative recall by allowing Darwin and Wallace to interact with each other.

All the repertoires of Darwin II are made by connecting together groups that have a common logical structure, summarized in Figure 4.7. There are two classes of input connections. Specific connections (upper left) may come from the input array or from groups in the same or other repertoires. The sources of all these connections are specified by lists, the construction of which differs from one repertoire to another. There are also short-range inhibitory connections (lower left) having a function corresponding to lateral inhibition in neural nets. These connections are geometrically specified and nonspecific. The level of activity at each input connection, if it exceeds a necessary threshold level, is multiplied by a weight corresponding to the strength of its particular synapse. The weight establishes how important the particular input is in determining the overall response of the entire group. The weighted inputs are all combined by adding the contributions from the excitatory inputs and subtracting the contributions from the inhibitory inputs, as suggested in Figure 4.7. The combined input must exceed a second excitatory or inhibitory threshold to have any effect on the group's activity. If not, the previous level of activity simply undergoes exponential decay. In either case, a varying amount of noise is added to the response of the group by analogy with the noise found in real neuronal networks. The final response obtained by combining all of these terms is made available to whatever other groups may be connected to this one (arrows at right).

To the extent that different groups are constructed with similar input connection lists and connection strengths, repertoires of these groups will have the required degeneracy. The specificities of the groups are implicit in their connection strengths—the best response is obtained when the most active inputs are connected to synapses with high connection strengths. The way these connection strengths are changed during the course of selection is suggested at the top of Figure 4.7 for a typical connection. There are many possible ways to formulate the rules for changing connection strengths. Fairly typical is this one: if a particular group responds strongly, and the input to one of its synapses is simultaneously active, the strength of that synapse should be strengthened so that later applications of the same input will give a still stronger response. Darwin II permits this so-called Hebb rule (Hebb 1949) or any of 80 other possible rules to be chosen for each type of connection. As the figure suggests, these rules have in common that the change in connection strength depends on the activity of either or both of the pre- and postsynaptic activities and on the pre-existing value of the connection strength but not on any other variables. Within limits, it is not too important exactly what rule is used, as long as it is recognized that connections must be able to decrease in strength as well as increase—otherwise, all synapses are eventually driven to maximum strength, and the system ceases to show selectivity.

Results

In examining the results obtained with Darwin II, we will be looking for the following criteria of success: (1) in Darwin, the generation of individual representations—that is, unique responses to each different stimulus and the same response to repeated presentations of the same stimulus, but stronger; (2) in Wallace, the generation of class representations—that is, similar responses to different stimuli having common class characteristics. In the complete system, these individual and class representations must interact to give associative recall of different stimuli in a common class.

Figure 4.8 shows the responses of the individual repertoires under conditions in which the cross connections between Darwin and Wallace are not functioning. It can be seen that the **R** responses are topographic, generally resembling the stimulus letters except for some occasional noise responses. The responses in **R-of-R**, as expected, are individual and idiosyncratic, and not at all topographic because features from different parts of **R** are being correlated. The responses to the two A's appear to be no more similar than the response to an A is to the response to an X, although statistics do show a somewhat greater degree of similarity, as we shall see. The situation in Wallace is entirely different. The responses in R_M are very similar for the two stimuli that are in the same class (the two A's), and are not at all sensitive to the idiosyncratic features of each letter. Moreover, the response is independent of any rotation or translation of a letter.

When selection is allowed to occur, in which synaptic strengths are modified in accord with an amplification rule, the responses become more specific, as shown in Figure 4.9. These response frequency histograms show the distribution of responses when a particular stimulus is first presented (a) and again after it has been presented for some time (b). Initially, most groups respond weakly, but these are a few groups that happen to respond very well. After selection, more good responders have been recruited from among the groups that earlier had been medium-strength responders, but large numbers of nonresponders remain available, possibly to respond to other stimuli not used in this experiment.

This behavior exemplifies *recognition;* that is, an enhanced response to a stimulus after it has been experienced before. If this same type of experiment is done with several different kinds of stimuli, the system displays *classification.* The similarity of the responses of a repertoire to any two letters can be measured by counting the number of groups that respond to both of the letters and dividing it by the number that would have been obtained by chance if the same total number of responding groups were distributed evenly over the repertoire. Classification can be assessed by examining ratios in which these similarity measures, obtained for pairs of letters in the same class, are divided by the corresponding similarity measures for pairs of letters in different classes. The class memberships of the different letters used are specified by the experimenter and are not available to Darwin II. In Table 4.1, we present values of this ratio for **R-of-R** and for R_M at the beginning of a typical experiment and after selection had been allowed to progress for three presentations of each stimulus. It can be seen that for these stimuli, groups in Wallace (R_M) were 91 times more likely to respond to both of two stimuli if they were in the same class than if they were in different classes. The system thus classifies by giving similar responses to different letters of the same kind. As suggested earlier, a minimal amount of classification is also seen in Darwin, as suggested by the 1.21 initial ratio. After selective modification of the connection strengths, the classification gets better, even though there is no feedback from the environment that would permit the system to "learn" which responses are "correct."

These results can be extended to stimuli not included in the training set (the set presented during selection) to yield *generalization* in both R_M and **R-of-R**. In R_M, this is a direct consequence of the class-responding characteristics of that repertoire, but in **R-of-R**, it is not, and, in fact, can be obtained only if re-entrant connections from R_M to **R-of-R** and within **R-of-R** are present. These connections permit R_M to influence the activity of **R-of-R,** supporting common patterns in the responses of disparate stimuli that happen to have similarities in their R_M responses by virtue of their common class membership. As Table 4.2 shows, in one particular experiment, the ratio of similarity of intraclass responses to interclass responses in **R-of-R** for a test set of letters not previously presented was 6.10 after selection based on other letters of the same kind (the training set), whereas it had been only 1.77 initially. The results for a control set of unrelated letters show that this effect is specific, and not due to a general increase in similarity of response to all stimuli.

These results indicate that the Darwin and Wallace networks separately fulfill the design criteria we have outlined. A final experiment demonstrates how the re-entrant connections between them work to give associative recall of stimuli that the system places in the same class by virtue of similar responses to them in the Wallace network.

The setup of the system for the association experiment is shown in Figure 4.10. Just two stimuli are used, an X and a +, chosen because their responses in **R-of-R** are quite different while their responses in R_M are very similar (because each consists of a pair of lines crossing near their centers). When the X is first presented (Figure 4.10, left panel), **R** (center left) gives the expected topographic response; **R-of-R** gives a unique pattern characteristic of that stimulus (for clarity, only a single group is shown responding in the figure), with stimulation via Pathways 1. At the same time, a trace occurs in Wallace, eliciting an appropriate pattern of response in R_M. Cross connections are present in both directions between **R-of-R** and R_M; connections that happen to join responding groups in the two repertoires are strengthened by the normal modification procedure (Pathways 2). In the center panel of Figure 4.10, the X is removed and a + is presented. The groups active in the response to the X are now no longer active (open circles), although the connections between them remain strengthened (solid lines). New groups in **R** and **R-of-R** become active in response to the + (filled circles); connections between these groups are strengthened as before (dashed lines, Pathways 1). In Wallace, the trace pattern is the same as for the X, eliciting a response in R_M very similar to that obtained before. Connections between the + responding groups in **R-of-R** and these same responders in R_M are therefore strengthened (Pathways 2). An indirect associative pathway is thus established via R_M between groups involved in the two patterns of response in **R-of-R**.

The third panel in Figure 4.10 shows how the association is tested. The trace mechanism is turned off so that the association will be based entirely on past experience with the stimuli and not on any immediate correlation occurring during the test. When the X is presented under these conditions, **R** and **R-of-R** give responses very similar to those obtained originally with the X. R_M receives input only from **R-of-R** via the previously strengthened Pathways 2, eliciting the common pattern of response appropriate to both the X and the +. Pathways 3 then permit R_M to stimulate in **R-of-R** the pattern originally associated with the second stimulus, the +, even though the + is not then present on the input array. Depending on the time constants chosen, this associated response can occur together with the X response, or later.

Results of a typical experiment of this type are shown in Figure 4.11. Responses are plotted as a function of time for a number of individual **R-of-R** groups. On the left (a) are seen the results when the test stimulus was the X. The groups at the top are ones that responded to the X during the first presentation of that letter, and the groups at the bottom are ones that responded to the + during the first presentation of that symbol. As expected, these + responders do not respond immediately when the X is presented. After four cycles of stimulation, **R** repertoire is switched off (arrow) so that it no longer dominates the **R-of-R** response, and there is now no outside input to either Darwin or Wallace. Under these conditions, the response of some of the X responders (top) begins to decay away (see, e.g., groups 77, 85, and 91), while some of the + responders now become active as a result of the stimulation they receive through the re-entrant connections from R_M (see, e.g., groups 28, 34, and 95). Thus, the system, presented with an X, recalls elements of the response proper to the +; these two stimuli have become associated.

A separate, reciprocal experiment is shown in Figure 4.11b, where the same groups are plotted for the case where the + is the test stimulus. Now it is the + responders (top) that begin responding immediately, but decay when **R** is switched off (arrow), and it is the X responders (bottom) that come up in associative recall when the stimulus is removed. Thus, the association is bidirectional—the + is associated with the X and the X is associated with the +. No association occurs if the original stimuli have no common response in Wallace (e.g., an A and an X).

In this experiment, letters are associated in Wallace in a trivial way, because their responses are very similar to begin with. In Darwin, however, their responses are not the same, but have individual character. The association obtained in Darwin in this experiment could not have been obtained without Wallace, but, at the same time, it goes beyond what Wallace could have done alone, because in Darwin, the individual character of the responses to the two stimuli is preserved. This example illustrates some of the features we suggested earlier for human perception. It is intended to clarify our assertion that pattern recognition is enhanced by the interaction of two basic methods of category formation: one based on the use of exemplars, the other based on common class features.

Summary

The results we have presented demonstrate that a network based on a selective principle can function in the absence of forced learning or an a priori program to give recognition, classification, generalization, and association. While Darwin II is not a model of any actual nervous system, it does set out to solve one of the same problems that evolution had to solve—the need to form categories in a bottom-up manner from information in the environment, without incorporating the assumptions of any particular observer. The key features of the model that make this possible are: (1) Darwin II incorporates selective networks whose initial specificities enable them to respond without instruction to unfamiliar stimuli; (2) degeneracy provides multiple possibilities of response to any one stimulus, at the same time providing functional redundancy against component failure; (3) the output of Darwin II is a pattern of response, making use of the simultaneous responses of multiple degenerate groups to avoid the need for very high specificity and the combinatorial disaster that would imply; (4) re-entry within individual networks vitiates the limitations described by Minsky and Papert (1969) for a class of perceptual automata lacking such connections, and (5) re-entry between intercommunicating networks with different functions gives rise to new functions, such as association, that either one alone could not display. The two kinds of network are roughly analogous to the two kinds of category formation that people use (Smith and Medin 1981): Darwin, corresponding to the exemplar description of categories, and Wallace, corresponding to the probabilistic matching description of categories.

These principles lead to a new class of pattern-recognizing machine of which Darwin II is just an example. There are a number of obvious extensions to this work that we are pursuing. These include giving Darwin II the capability to deal with stimuli that are in motion, an ability that probably precedes the ability of biological organisms to deal with stationary stimuli (Ullman 1979), giving it the capability to deal with multiple stimulus objects through some form of attentional mechanism, and giving it a means to respond directly and to receive feedback from the world so that it can learn conventionally. Already, however, we have shown that a working pattern-recognition automaton can be built based on a selective principle. This development promises ultimately to show us how to build recognizing machines without programs and to provide a sound basis for the study of both natural and artificial intelligence.

Bibliography

Dennis, I., J.A. Hampton, and S.E.G. Lea, 1973: New problems in concept formation, *Nature* 243:101–102.

Edelman, G.M., 1978: Group selection and phasic reentrant signalling—a theory of higher brain function, in G.M. Edelman and V.B. Mountcastle (eds.), *The Mindful Brain*, MIT Press, Cambridge, Mass., 51–100.

Edelman, G.M., and L. Finkel, 1984: Neuronal group selection in the cerebral cortex, in G.M. Edelman, W.M. Cowan, and W.E. Gall (eds.), *Dynamic Aspects of Neocortical Function*, John Wiley & Sons, New York, 653–695.

Edelman, G.M., and G.N. Reeke, Jr., 1982: Selective networks capable of representative transformations, limited generalizations, and associative memory, *Proceedings of the National Academy of Sciences USA* 79:2091–2095.

Hebb, D.O., 1949: *The Organization of Behavior*, John Wiley & Sons, New York.

Lashley, K., 1950: In search of the engram, in *Physiological Mechanisms in Animal Behavior* (Society of Experimental Biology Symposium No. 4):454–482, Academic Press, New York.

Minsky, M., and S. Papert, 1969: *Perceptrons—An Introduction to Computational Geometry*, MIT Press, Cambridge, Mass.

Smith, E.E., and D.L. Medin, 1981: *Categories and Concepts*, Harvard University Press, Cambridge, Mass.

Ullman, S., 1979: *The Interpretation of Visual Motion*, MIT Press, Cambridge, Mass.

FIGURE 4.1 Fragment from Maurits C. Escher's woodcut
<u>Metamorphosis II</u>, illustrating the importance of context in
perception. (c) M.C. Escher Heirs, c/o Cordon Art BV Baarn
(Holland). Reprinted with permission.

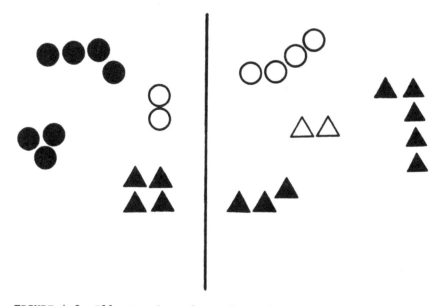

FIGURE 4.2 Illustration of a polymorphous rule for set
membership. Members of the set (left) have any two of the
properties roundness, solid color, or bilateral symmetry.
Nonmembers (right) have only one of these properties.
Reprinted by permission from <u>Nature</u>, Vol. 243, pp. 101-102.
Copyright (c) 1973 Macmillan Journals Limited.

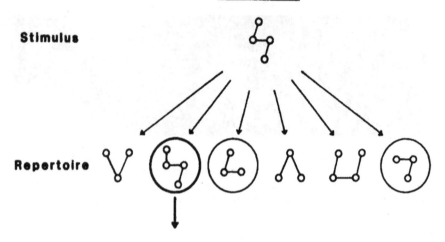

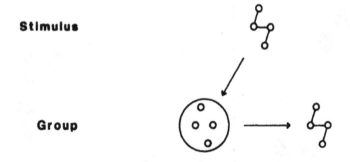

FIGURE 4.3 Schematic representation of the neuronal group selection theory (top) contrasted with information-processing or instructional theories (bottom). Circles connected with lines represent specific characteristics of stimuli and the matching characteristics of groups that respond to them, not any particular neuroanatomy. Groups whose specificity matches a given stimulus more or less well respond (groups surrounded with larger circles); only those groups that respond best (heavy large circle) are modified for stronger future response (heavy arrow). In instructional theories, groups have no particular initial specificities (group without joining lines, bottom); specificity is dictated by interaction with stimuli.

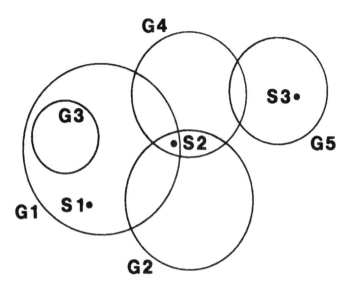

FIGURE 4.4 Degeneracy. Stimuli are represented by heavy dots labeled S1, S2, etc. Recognizing groups are represented by circles labeled G1, G2, etc. The size of each circle suggests the range of stimuli to which a group can respond. Overlapping circles suggest how multiple groups from a degenerate repertoire can respond more or less well to the same stimulus.

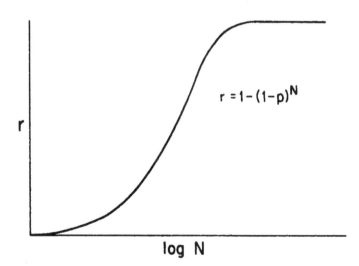

FIGURE 4.5 Response as a function of repertoire size. Assuming a constant, independent probability p that any group will respond to any one stimulus, the function r gives the probability that some one or more groups in a repertoire of N groups will respond to any one stimulus.

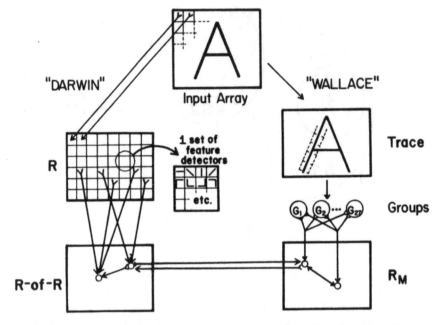

FIGURE 4.6 Simplified plan of construction for Darwin II

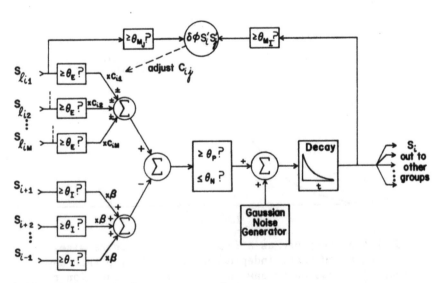

FIGURE 4.7 Logical structure of a group

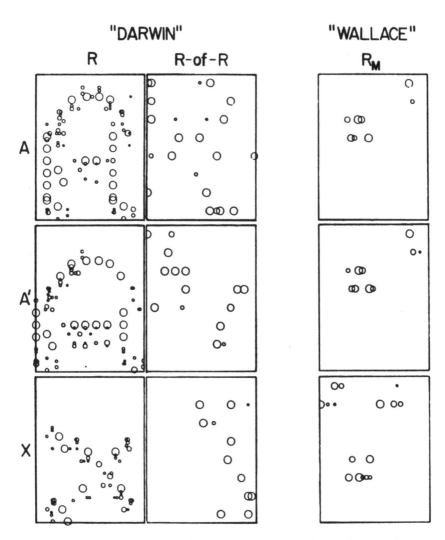

FIGURE 4.8 Responses of individual repertoires (R, R-of-R, and R_M; names at top) to a tall, narrow A (top row), a lower, wider A (middle), and an X (bottom). Circles represent groups responding at 0.5 (small circles) or more of maximum response (large circles). Groups responding at less than half of maximal response are not shown, leaving blank areas in the repertoire plots.

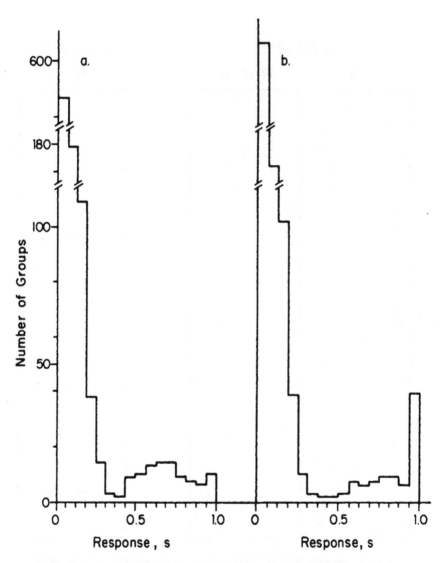

FIGURE 4.9 Response frequency histograms. (a) Initial
response to a novel stimulus. (b) Response to the same
stimulus after selection has proceeded for a time. Abcis-
sae: response levels s, expressed as a fraction of maximal
response. Ordinates: numbers of groups responding at level
s.

TABLE 4.1 Classification in Darwin II[a]

| | Repertoire | |
| | Darwin | Wallace |
Time Tested	[R-of-R]	[R_M]
Initially	1.21	90.93
After selection[b]	1.41	241.30

[a]Repertoire sizes: R, 3,840 groups; others, 4,096 groups. Total connections: 368,840; no Darwin-Wallace connections. Stimuli used: 16 letters, 4 each of 4 kinds. Quantity shown is ratio of number of groups responding to two stimuli in same class to number of groups responding to two stimuli in different classes, corrected for numbers that would respond in each case by chance alone.

[b]Each of four stimuli was presented for eight cycles, then the entire set was repeated three times.

TABLE 4.2 Generalization in R-of-R[a]

Stimuli	Intraclass / Chance	Interclass / Chance	Intraclass / Interclass
Initially			
Training set	2.09	0.72	0.90
Test set	2.89	1.63	1.77
Control set	--	1.96	--
After selection[b]			
Test set	6.10	1.00	6.10
Control set	--	1.00	--

[a]Repertoire sizes: R, 3,840 groups; others, 1,024 groups. Connections to each R-of-R group: 96 from R, 64 from R-of-R, 128 from R_M. Stimuli used: 16 letters, 4 each from 4 classes.

[b]Each of 16 stimuli was presented for 4 cycles, then entire set was repeated 4 times.

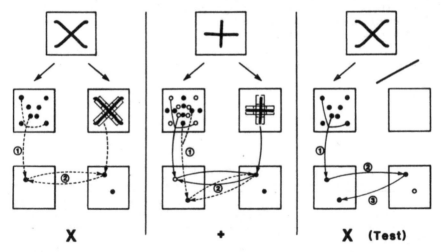

FIGURE 4.10 Schematic views of Darwin II showing three
stages in an associative recall experiment. Filled circles
represent active groups; open circles, inactive groups.
Solid lines between groups represent connections selectively
strengthened; dashed lines represent connections activated
for the first time. Numerals enclosed in circles label
pathways that are activated at successive times.

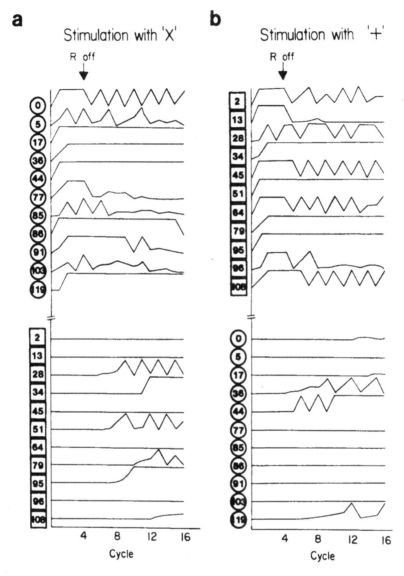

FIGURE 4.11 Responses of individual R-of-R groups in an associative recall test. Response of each group is plotted as a function of time, measured in cycles of the model (scales at bottom). Each group is labeled by its serial number in the repertoire. A serial number is enclosed in a circle if the group responded initially to an X or is enclosed in a box if the group responded initially to a +. Arrows (top) indicate time (after fourth cycle) at which all input to a system was cut off. (a) Stimulation with an X. (b) Stimulation with a +.

5. Biologically Motivated Machine Intelligence

Biological systems routinely solve problems involving pattern recognition and feature extraction. Such problems do not appear to admit similarly routine algorithmic solutions; the power of biological systems in this regard apparently arises from nonalgorithmic dynamics. It is our intention to explore and to develop principles of functional characteristics of natural (biological) intelligence, with a view to their realization through an appropriate automaton model.

A basic characteristic of biological systems is that they tend to evolve, through modes of information processing, toward "suitable" modes most appropriate to their environment. The dynamics of this process has been captured and indirectly exploited in simulated annealing, as used in attacking a variety of optimization problems (Metropolis et al. 1953; Kirkpatrick, Gelatt, and Vecchi 1983; Geman and Geman 1984; Hinton et al. 1985). The prototypic example of an annealing process consists of a gas molecule confined in a potential well in which the goal is the location of a global potential minimum. If the ambient temperature is lowered sufficiently slowly, the molecule will become trapped in the global minimum with a probability arbitrarily close to one. It is the random behavior of the molecule, "thermal noise," which accounts for its ability to escape from local minima during the cooling process. Simulated annealing then entails simulation of these dynamics in the solution of combinatorically large-scale minimization problems such as the "travelling salesperson" problem. The essential role played by random noise in such techniques places them outside the realm of algorithmic strategies.

Reprinted with permission from the *SIGART Newsletter*, No. 95, p. 31, January 1986, Association for Computing Machinery, New York.

The authors carry this trend one step further in a discussion of evolutionary learning systems (Hastings and Waner 1985). Briefly, an evolutionary learning system is a formal dynamical system in which the states correspond to modes of information processing, while the suitability of each state is measured by a potential function, the most desirable states possessing least potential. The dynamics of such a system is determined by an annealing process, so that desirable modes are attained by a gradual lowering of the amount of thermal noise. Further, the potential energy function depends on the environment, so that it is the environment which indirectly determines the equilibria and evolution of the system. We refer to this indirect process of control as *soft programming*. An evolutionary learning system may then be thought of as a dynamical system which behaves according to three principles: *ergodicity*—the principle of chaotic behavior; *annealing*—the regulation of thermal noise by means of (local) lowering of ambient temperature; and *soft programming*—the indirect control of the evolution of the system by the environment. The authors further describe how these principles may be implemented through a suitably designed stochastic neural network.

Principles of Machine Intelligence

Whereas evolutionary learning is an essential attribute of biological intelligence, the ability to learn is by no means sufficient to account for pattern recognition, feature extraction, and, in the case of "sentient" intelligence, self-reference. We, therefore, wish to expand on these principles in order to account for such specific abilities.

For purpose of convenience, we think of an evolutionary learning system as a special kind of "history-dependent stochastic automaton."

By a *history-dependent stochastic automaton (HDSA)*, we mean a tuple (I,S,O,R,P), where:

I is a set (of external inputs), including a null state;
S is a (possibly infinite) set of internal states, including
an initial state;
O is a set (of outputs);
R is a function from the set of states to the set of outputs; and
P is a function which assigns to each *history*.

$$H_r = (s_0,i;0,s_1,i_1, \ldots ,s_r,i_r)$$

with s_j in s and i_j in I, a probability measure $P(H_r)$ on the set of states.

We remark that if P depends only on the "current" data (s_r, i_r), we have a classical stochastic automaton. Notice also that the environment of such an automaton may be thought of as consisting of finite sequences of input symbols, or equivalently, as elements of the free monoid M on the set of input symbols. One can view a HDSA as part of a dynamical system. In this context, a mode m of information processing is a specified behavior as a function of input sequences. We then postulate that, in order to reflect the behavior of a biological system, a HDSA should realize the following principles.

Principle I—Evolutionary Learning

A suitably designed automaton incorporates the principles of ergodicity, annealing, and soft programming, as previously discussed. (Notice that, by definition, HDSAs are ergodic systems, while the remaining principles of evolutionary learning may be incorporated as features of the history dependence together with suitable feedback. This is discussed further later.)

Principle II—Feature Extraction

In the context of our simple input environment, a "feature" will be understood as some finite sequence of symbols. Formally, a feature is then an element f of the free monoid M, and two finite input strings have the common feature f if the symbols of f occur in order as a (not necessarily contiguous) subsequence of both strings. Operationally, we may regard the feature f as having been extracted when the automaton responds similarly to inputs with common feature f. Thus, the automaton should, in response to a suitable "instruction" input sequence, return via its output indication that it has distinguished input sequences including f from those without it.

Even the simple environment posed here is inherently hierarchical; the inclusion $f \rightarrow f'$ of subsequences of symbols endows the monoid M with a natural partial ordering, and thus with an inherent hierarchical structure. Our third principle asserts that the automaton must possess the ability to form internal hierarchies which reflect this structure.

Principle III—Hierarchical Learning

An intelligent automaton should learn to form hierarchical structures deriving from the features it extracts. Thus, not only should it learn to extract features on different hierachical levels (as determined by the partial ordering referred to previously), but it should learn the hierarchical structure, as evidenced by its outputs.

The ability to associate elements of M with each other should not be dependent on their possessing an inherent common feature. For example, we associate knives and plates not via any inherent geometric similarity, but rather through our experience of their association. This phenomenon motivates the next principle.

Principle IV—Associative Learning

Elements f and f′, which are presented "sufficiently often" in "close" temporal proximity, should be associated as common features. Again, hierarchies are inherent in associated input sequences: if f and f′ are associated, and if $f_l \geq$ f′, then one may view f_l as $\geq$ f′. Associative learning is thus required to include the learning of the derived hierarchies.

Perhaps the most difficult requirement for an intelligent machine is that it possess the ability to respond to its own dynamics in a meaningful way.

Principle V—Self-reference

An intelligent automaton should include in its internal structure an infinite hierarchy of machines, each responding to the dynamics of those below in such a way as to realize Principles I through III vis-à-vis these dynamics.

Roughly speaking, this principle models the natural hierarchy in "thinking," "thinking about one's thoughts," and so on. It also results in the ability of an automaton endowed with this ability to form abstract hierarchies based upon operations which form part of its acquired dynamics. For example, a machine which has learned the notions of additive and multiplicative identities could now extract the common internal dynamic of identity operations and recognize it as a feature in itself.

Implementation

Previously (Hastings and Waner 1985), the authors described how the principles of evolutionary learning may be embodied in a simple stochastic neural network. Although the simplicity of design restricts the variety of possible modes of information processing, the net is able to realize the three principles of evolutionary learning. Later (Waner and Hastings, Chapter 7, this volume), we demonstrated how a suitable enrichment of the net's structure can lead to the learning of arbitrary logical modes of information processing. We present here, in outline, a somewhat modified version of the original (simple) network, and give plausibility arguments which amount to asserting that the (modified) network is capable of realizing all five principles.

One starts with a rectangular lattice in which the nodes represent (locations of) formal neurons. Each neuron is connected to all others in a suitable neighborhood, and possesses an activity of 0 and 1. If the activity of a neuron is 1, then the neuron fires to all its neighbors. Conduction along edges is stochastic; the probability that an edge will conduct being given by 1 − (conduction threshold). Each conducted impulse activates its target. Upon firing, a neuron becomes deactivated (unless it receives an incoming impulse from a neighbor). At each time step, all the firings take place in parallel.

In order to achieve local cooling, thereby realizing the annealing principle, the conduction threshold of an edge which conducts is lowered (upon conduction). The amount by which such thresholds are lowered is determined by a suitable metric on the unit interval; the closer a threshold is to 0, the smaller the decrements. Edges which fail to conduct are correspondingly increased in order to prevent explosiveness in overall activity. Superimposed on this is a tendency for edges which are inactive to decay toward a global base value, thus effecting local stochastic "reheating." Previously (Hastings and Waner 1985), we showed how such a net can learn to form directed graphs through soft programming, while in the later paper, we showed how the inclusion of stochastic bit masking in the structure permits the net, in principle, to learn arbitrarily complex modes of information processing.

As presently structured, the net is, as we have indicated, capable of realizing Principle I. In order to realize Principle IV (associative learning), we modify the threshold-lowering regimen slightly: assume that neuron n is presently active, and fires to neighbors $a_1, a_2, \ldots a_r$, and that, in addition, neighbors $b_1, b_2, \ldots b_s$ are activated at the conclusion of the time step (either via conduction from other neurons or via external activation). Then, the lowering regimen proceeds as before, but threatens the edges from n to the b as also having conducted. It then is not difficult to see that if external stimuli are presented more-or-less simultaneously at two "reasonable" close sites n_1 and n_2 repetitively, that paths will develop from n_1 to n_2 as a direct result of this modification. Principle IV is then essentially an elaboration of this simple mechanism.

Once the net is capable of associative learning, Principles II and III are automatic: both the extraction of common features and the formation of hierarchies emerge as natural consequences of associative learning in the network. Principle V is somewhat serendipitous: since the inputs to the net are realized as the stimulation of neuron sites, and since the subsequent dynamics do not distinguish externally stimulated sites from internally stimulated ones, the sequences of active sites in the net itself may be thought of as input sequence, and their secondary effects on the dynamics as a response to these (primary) dynamics. Thus, the net may be thought of as responding to its own dynamics in exactly the same way as it responds to outside stimuli. This property may be thought of as a form of structural self-similarity, in which successive scales represent levels in a hierarchy of references. In the limit of an infinite net, one then has true self-similarity, and, thus, true self-reference.

Bibliography

Geman, S., and D. Geman, 1984: Stochastic relaxation, Gibbs distributions, and the Bayesian restoration of images, *IEEE Transactions on Pattern Analysis and Machine Intelligence* 6:721–741.

Hastings, H.M., and S. Waner, 1985: Principles of evolutionary learning—design for a stochastic neural network, *BioSystems* 18:105–109.

Hinton, G.F., T.J. Sejnowski, and D.H. Ackley, 1985: Boltzmann machines—constraint satisfaction networks that learn, *Cognitive Science* 9:147–164.

Kirkpatrick, S., C.D. Gelatt, Jr., and M.P. Vecchi, 1983: Optimization by simulated annealing, *Science* 220:671–680.

Metropolis, N., A. Rosenbluth, A. Teller, and E. Teller, 1953: Equations of state calculations by fast computing machines, *Journal of Chemical Physics* 21:1087–1091.

6. Concurrent Computation and Models of Biological Information Processing

Biological Information-Processing Systems

The ultimate laws that govern the operation of biological information-processing systems and electronic information-processing systems are surely identical, but the presently known examples of these two types of computing engines are strikingly different in their capabilities and limitations. By biological standards of performance, electronic information-processing systems are very fast and error free, but electronic information-processing systems are far surpassed by their animate cousins in their ability to recognize and categorize patterns, to reason, and to learn.

This inequality of performance is certain to be diminished, for the rate of evolution (so it must be termed!) of microelectronic means for storing and processing information is very much greater than the rate of evolution of organisms—particularly of complex organisms. If this year's chip can store 1 million bits of information, next year's will probably store nearly twice as many; if this year's processor can execute 1 million instructions per second, next year's will probably execute one-third again as many. The powers of electronic information-processing systems will undoubtedly increase by many orders of magnitude during the next century. But next year's brain will neither store more bits nor execute more instructions per second than this year's model. These developments will create the opportunity to employ electronic machines for purposes that require intelligence. Thus, it is of considerable practical importance, as well as profound scientific significance, to understand how the human brain, a tangled network of undetermined and variable topology manufactured from slow and unreliable components by willing but unskilled labor, can perform so admirably on the subtle problems of pattern categorization and associative retrieval, and on learning and making decisions in the face of uncertainty. There must be something special about the design of the neural network considered as an information-processing *system* that enables the emergence of these powerful collective properties from aggregates of simple components.

Part of the power of the brain as a computing machine certainly derives from its sheer size: there may be as many as 10^{12} neurons in this powerful, air-cooled portable computer, exquisitely packaged in a 1400-cubic-centimer container sporting the latest hairdo and consuming only about 15 watts of power for its operation. The system is also equipped with high bandwidth input channels: the color-sensitive photoreceptors in the retina of the eye respond to about 8 bits of illuminant intensity variation and project onto as many as 1 million sites in the lateral geniculate nucleus of the cortex. Thus, as many as 10^9 bits per second of visual information input can be processed by the visual cortex while we listen to a Bach violin partita. Considering the silly things that people are apt to say, it is perhaps fortunate that the bandwidth of the output channels is much smaller.

The enormously large number of neurons that constitute the computing elements of the brain is still not sufficient for it to deal with most problems of reasoning and recognition in a direct and comprehensive way. Even this remarkable computing machine is constrained, and from its constraints we can, perhaps, learn something about the way the brain is wired and about the principles of efficient design of information-processing systems in general.

Neurons typically are capable of changing state as frequently as 1000 times per second. Thus, their switching time is measured in milliseconds, whereas the reaction time of the organism as a whole is typically measured in tenths of a second. Hence, howsoever information may be distributed throughout the nodes and interconnections of the neural network, there can only be about 100 steps, 100 ticks of the "clock" in the "program" that leads from a visual or acoustic or other sensory input to the signals that initiate motor responses to them.

The brain is not a synchronous computer: in reality, there is no clock signalling global changes of state. But there is an average switching time, and this way of thinking about the succession of states of the brain will not lead us astray, at least insofar as the question of overall performance limitations is concerned.

In 100 program steps, a neuron can influence others that are no more than 100 network links distant from it. Thus, thought processes are local in the sense that they cannot propagate far from their source, where "distance" is measured by the number of links that a signal traverses. Unless the connectivity of the neural network is immense, the signals transmitted by a given neuron will be communicated (in combination with the outputs of other neurons) to at most a small fraction of the totality of neurons.

It was Helmholtz who discovered that neural signals are propagated with finite speed by measuring how fast they travel. Although the speed of propagation of signals through the neural network varies to a considerable extent depending upon the diameter of the neuron through which it propagates and whether the neuron is myelinated, it is not unlike the speed of light in two ways: information cannot be transmitted through the network at a greater speed, and computational causality is constrained to the analog of the "forward light cones" of the Special Theory of Relativity. Thoughts—i.e., computations—that occur in one part of the brain may not be able to influence those that occur in another part, at least for a considerable period of time. If, for example, the neurons of the brain were connected in the pattern of a 1-dimensional chain, then a given neuron could, within the standard 100 ticks of the global clock, signal the news of its stimulation to at most 200 others.

One may say that the brain as a whole must have no idea what its parts are up to; they function quasi-autonomously. This argument, based on considerations of the connectivity and channel capacity of the network, leads naturally to the point of view of thought processes that Minsky has called the "society of mind," and perhaps provides something deeper than a phenomenological argument in support of it.

Consider, for instance, a 1-million-pixel image of a part of a visual scene. If neurons corresponded to individual pixels at some level of representation of the image, and if two neurons were connected if and only if they corresponded to nearest neighbor pixels in the image, then lateral interactions would be confined to a region having a radius of about 100 pixels, corresponding to about 3% of the image.

Nearest neighbor connections of N nodes in a 2-dimensional rectangular array requires approximately 2N connections. Were each node directly connected to every other one, then approximately $(1/2)N^2$ connections would be required: 5×10^{11} for a 1-million-pixel "retina" and 5×10^{23} for the whole brain. The former would be difficult to accomplish; the latter seems both electronically and biologically impossible.

Let us try to arrive at an estimate of the connectivity of the neural network, in order to assess the types of network topology that might be possible. Our estimate will necessarily be imprecise, but we hope that it will serve to stimulate some new ideas and approaches to this important question.

A direct anatomical attack on this problem will fail. A single neuron may have tens of thousands of synaptic connections to other neurons, although the average number may be two or three orders of magnitude smaller. Recent studies suggest that many of the synaptic junctions emanating from one neuron redundantly connect it to one other neuron. Some estimates suggest that almost all of the connections emanating from many-synapse neurons are redundant. The redundant ones probably should be combined into equivalence classes when we think about network connectivity.

We can estimate the number of active connections per neuron from an entirely different standpoint. Consider the neuron as a switch that can change state in approximately 1 millisecond (msec), and suppose that the average switching energy per switched connection is E_{sw}. If the average number of active connections (i.e., connections carrying signals) is C, then the energy required to change the states of the neuronal network is $C E_{sw}$ per switching cycle. Let us assume that this quantity is equal to the fraction r of the total energy E consumed by the brain in this time. Then

$$C = r \frac{E}{E_{sw}}$$

It is easy to estimate E. The brain of a young adult produces about 4.8 kilocalories per liter of oxygen consumed, and it consumes about 0.046 liters of oxygen per minute. This implies that the energy consumption of the brain is approximately 220 calories per minute, or equivalently, 15.3 watts. The lightbulb as the symbol of "the better idea" is not at all an inappropriate icon. It is surprising, however, to learn that a better idea need not necessarily be a very bright idea.

It will be convenient to express the power consumption of the brain in units that are more naturally related to information-processing magnitudes. We find that 15.3 watts $= 9.6 \times 10^{19}$ ev/sec, where ev denotes electron volts. Thus, we obtain the estimate $E = 9.6 \times 10^{16}$ ev per switching cycle for the energy consumption of the brain.

Let us assume that the fraction r of the power consumed by the brain that is devoted to switching neural circuits satisfies $r \leq 1/3$. Then, the number of active connections C per switching cycle will be constrained by the inequality

$$C \leq \frac{3.2 \times 10^{16}}{E_{sw}} \, ev$$

where E_{sw} is the average energy required to switch a neuron from one state to another.

An absolute lower bound for E_{sw} is given by kT, the background energy due to thermal agitation, where k denotes Boltzmann's constant and T is temperature in degrees Kelvin. At normal temperatures, $kT \simeq 0.026$ ev, whence $C \leq 1.3 \times 10^{18}$. In this case, the mean number of active connections per neuron per switching cycle would, assuming $N = 10^{12}$ neurons, be bounded by the number $C/N \leq 1.3 \times 10^{6}$. Although this ratio is much smaller than the ratio that corresponds to a network in which each neuron is connected to every other one (for which $C/N = 5 \times 10^{11}$), it is nevertheless a disconcertingly large number.

This thermal background estimate of the energy required to switch a neural "gate" is unrealistically small. Typical switching energies for gates built from transistors on integrated circuits are currently about 2×10^{5} ev, with the switch changing state in less than a microsecond. The energy consumption is 10^{7} times as great as thermal switching energies. If a neuron required a similar amount of energy to change state, then $E_{sw} \simeq 2 \times 10^{5}$ ev implies that the critical ratio C/N would be bounded by 0.14 connections per switching cycle per neuron. Biological information-processing networks are evidently more efficient than currently manufactured electronic information-processing systems from this viewpoint (see Table 6.1).

Single-channel currents in neurons involve the flow of 2 picoamperes at 90 millivolts for about 1.5 msec. The energy transferred is about 2×10^{3} ev. If we take this quantity as a rough estimate of the switching energy E_{sw} of a neuron, then $C \leq 1.6 \times 10^{13}$ and $C/N \leq 16$ active connections per switching cycle per neuron.

What kinds of network topology can be accommodated by this constraint? This degree of connectivity suffices to insure that the collection of $N = 10^{12}$ neurons can be arranged in a network consisting of a single connected component, but the connectivity of the network is significantly limited.

Although one should not place too much reliance on our estimate of the brain's connectivity, its possible significance is indicated by an interesting theorem of Erdös and Rényi (1959)[1] Erdös and Rényi are concerned with the likelihood that a randomly selected graph is connected. A *graph* is a finite set of (possibly labelled) points called *vertices* together with a collection of unordered pairs of points called *edges*. A graph is *connected* if it is possible to pass from any one of its points to any other by traversing a sequence of edges. There are

$$\binom{\frac{N}{2}}{C}$$

possible graphs with N labelled vertices and C edges. Suppose that one of these graphs is selected at random (relative to the uniform probability distribution). What is the probability that the selected graph is connected? Denote this probability by p(N,C). Erdös and Rényi introduce the quantity

$$C_a = \left[\frac{N}{2} \log N + N_a \right]$$

(where [x] denotes the greatest integer $\leq$ x and log is the natural logarithm) and prove that

$$p(a) = \lim_{N \to \infty} P(N,C_a) = \exp(-e^{-2a})$$

If a $>$ 0, it is very likely that the graph is connected, whereas if a $<$ 0, it is very unlikely that the graph is connected, as Table 6.2 indicates.

If N $=$ 10^{12}, which is our estimate of the number of neurons in the central nervous system, then[2] C_a/N $=$ 1/2 log N $+$ a $=$ 13.8 $+$ a.

We will identify C_a/N with our estimate of the mean number C/N of active connections per neuron. From C/N $=$ 16, we obtain a $=$ 2.2. Let us suppose that N $=$ 10^{12} is large enough so that $p(N,C_a)$ does not differ appreciably from its limiting value. Then p(2.2) $\simeq$ 0.988, which implies that the brain would probably be connected— but not with overwhelming likelihood—were it fabricated by a process of "random wiring."

Let us note that, in fact, the brain is just barely connected. The cerebral hemispheres are joined by the *corpus callosum* which, if severed, deprives the brain of the possibility of integrating information possessed by its severed components. The separated hemispheres can, however, lead virtually separate and effective existences which indicate both their mutual independence and the limited nature of the connections that normally join them. Those parts of the central nervous system that govern autonomic processes may also be only loosely coupled to the main portions of the neural network.

On the other hand, were the mean number of connections per neuron to fall below the critical constant 13.8, then the probability of the brain falling into many disconnected components would rapidly increase to near certainty. The problem of controlling a complex system whose components act autonomously and are incapable of communicating with one another is virtually intractable. Since the manufacture of the brain necessarily involves a considerable element of chance in determining exactly which pairs of neurons will or will not be connected, the theorem of Erdös and Rényi appears to provide compelling evidence that a brain must have at least an average of (1/2)logN connections per neuron in order to be viable and to preserve it from catastrophic failure as its individual neurons die.

The latter observation applies with equal force to massively parallel computing machines that are intended to operate despite the failure of many of their component parts. Although the network of interconnections may be anything but random at the outset, as time passes, it will increasingly approach randomness. This suggests that a parallel computer consisting of a large number N of processing nodes and C of connections among them should satisfy the inequality $C > (N/2)$ logN.

Let us consider several characteristic examples (cp. Broomell and Heath 1983; Haynes et al. 1982).

A balanced binary tree structure having a total of $10^{12} \simeq 2^{40}$ nodes would have a depth of 40. A message could be communicated between a pair of arbitrary nodes in not more than 78 switching cycles, which is within the 100 program step limit if most of the steps are merely concerned with forwarding information rather than performing calculations on data. Trees are unreliable, however: if a node or link fails, then the complete subtree having the node (or node on which the link terminates) as a root is disconnected from the remainder of the tree. The high reliability of the brain suggests that there are many redundant pathways, and hence that the basic hierarchical organization of information-processing structures is combined with path redundancy. This will further limit the "distance" that a signal can be propagated in 100 ticks of the clock.

For tree architectures, one finds $C = N - 1$, which rapidly falls behind (N/2)logN as N increases. This network is very fragile against the failure of a vertex or an edge; the remaining network has a significant probability of being disconnected.

In order to increase the effective utilization of the silicon from which the computer is fabricated, the designer must bring the level of activity of the memory into better balance with the activity of the processor and the communication network. This can be accomplished by causing the processor to act concurrently on a number of data elements stored in memory. The greater the number of data elements that can be simultaneously processed, the greater will be the efficiency of utilization of the fraction of the silicon surface devoted to memory.

Let us suppose that a computational problem admits decomposition into a highly parallelizable algorithm whose constituents can be processed concurrently. By associating individual processing elements with relatively small subsets of memory, the efficiency of utilization of the memory will be increased. But there are trade-offs. A powerful serial processing unit can execute only one instruction at a time, so the portions of its circuitry that correspond to alternative instructions will be idle at a given program step. It follows that manifold replication of powerful processing elements in order to increase the efficiency of memory utilization will decrease the efficiency of utilization of the silicon surface devoted to the processing elements. This argument implies that the replicated processing elements should be as simple as possible in order to decrease the fraction of idle processing circuitry as instructions are executed.

Thus, we are led to consider a computer architecture wherein large numbers of simple processors operate concurrently, each executing instructions on data drawn from a relatively small localized memory. In order to permit the processors to share instructions and data, and to modify their state depending upon the state of other processors, processors and memory should be able to exchange information by means of an extensive communication network. This is the essence of the Connection Machine™ computer design philosophy.

The Connection Machine™ ("CM") computer (Hillis 1981) is a family of parallel computing machines characterized by an unusually large number of very simple processing elements, each of which is associated with a small amount of local memory but has access to the full memory of the machine through a high-bandwidth routing network. Approximately 25% of the silicon surface in the CM computer is devoted to processing elements, and another 25% to the communication network which routes messages amongst the processors; memory accounts for the remaining 50%.

The CM operates in conjunction with a "host" processor and conventional magnetic disk mass storage peripheral devices. The host processor is a conventional mini-computer, such as a Symbolics LISP machine or a Digital Equipment Corporation VAX, which is used to synchronously transmit the stream of program instructions to a microcontroller, which, in turn, broadcasts them simultaneously at a high speed to all processing elements of the CM by means of a communication network configured as a tree.

Each processing element executes the instruction (in a possibly modified form depending on its current state) on data available in the processor's local memory. Thus, the CM belongs to the class of synchronous "single instruction multiple data" (SIMD) concurrent computer architectures, although the ability of the individual processors to modify the broadcast instruction depending upon their current states (just as each member of an orchestra "modifies" the instruction broadcast by the conductor depending upon the score for the individual instrument) provides some of the flexibility of "multiple instruction multiple data" (MIMD) architectures.

The CM can be thought of as a "smart memory": each cluster of memory locations has associated with it a small processing element. The CM is attached to the host processor through its memory, and the latter can access the CM memory elements individually in order to load data into and out of the CM. The CM processing elements and associated local memory are arranged in a 2-dimensional rectangular communication array. One edge of this array is connected in parallel to a system of magnetic disk memory units, thereby providing a high bandwidth input/output interface. The CM memory can be rapidly loaded or read by simultaneously shifting data in parallel along the columns of the rectangular array. This communication network also provides immediate communication between a processing element and each of its four nearest neighbors, which is of particular value for image processing and other applications which have a natural 2-dimensional "nearest neighbor" topology.

A rectangular array communication network does not provide a particularly efficient means for transmitting messages from one node to another, because the maximum message delivery time is proportional to the perimeter of the array, and hence increases at least as rapidly as the square root of the number of processors. For machines employing tens of thousands or millions of processors, the time required to transmit a message can be prohibitive. A communication network having a greater degree of connectivity is essential.

In addition to the rectangular array, the CM employs a boolean hypercube network to reduce message delivery time. The machine contains a total of $N = 2^d + p$ processing elements, with 2^p processors arranged on a single integrated circuit chip which also contains circuitry for routing messages through the communication network. There are a total of 2^d integrated circuit chips which, along with their associated local memory, are connected in a packet-switched, d-dimensional boolean cube communication network. Thus, a message can be sent from any one of the 2^d chips to any other one in time that is proportional to d. The 2^p processors on each chip are connected by a cross-bar switch so that it requires only one communication cycle for a message from a processor to be routed off-chip or for an arriving message to be received by the addressed processor in the absence of collisions.

The communication circuits contain provisions for automatically re-routing messages in the event of collisions.

The custom-designed integrated circuit containing 2^p processors and the routing circuitry is accompanied by 2^m bits of local memory. This combination is replicated throughout the machine 2^d times. It provides an unusually high degree of symmetry of fabrication, which has beneficial consequences for diagnostic testing and maintenance of the machine. The total amount of memory is $M = 2^d + m$ bits.

The theme of symmetry and simplicity in the design of the CM is carried down to the detailed structure of the processing elements themselves. Each processing element is a bit-serial computer (i.e., it operates on the data stream one bit at a time). The processor can compute any of the 2^{16} boolean functions of 3 inputs and 2 outputs. The inputs can be thought of as a pair of operand bits, augmented by a third bit that characterizes the current state of the processing element. One of the outputs can be thought of as a generalization of the "carry" bit when the operation is addition.

Currently available technology could be employed to fabricate a CM with $N = 2^{20} \simeq 10^6$ processing elements and $M = 2^{34}$ bits of memory. Advanced technology concepts suggest that machines having much larger numbers of processors could be built. At that stage, it should be possible to simulate models of neural network performance in considerable detail.

At the present time, a CM prototype is being fabricated to test the architectural concepts that underlie its design and to explore applications that require the largest scale of computing power. The prototype consists of $N = 2^{16} \simeq 64,000$ processing elements with $2^4 = 16$ processing elements arranged on a single chip which also incorporates the routing circuitry for the communication network. The chips are arranged in a $d = 12$-dimensional boolean hypercube packet-switched communication network. Each processor/router chip is associated with 2^{16} bits of memory; there are a total of $M = 2^{28}$ bits $= 32$ MBytes of memory in the machine. The prototype CM operates at a nominal maximum speed of 1000 MIPS (1 MIPS $= 1$ million instructions per second) for fixed-point additions of 32-bit words. The high-bandwidth input/output channel operates through a 128×512 rectangular array network, and has a capacity of 500 Mbits per second.

The CM prototype can be partitioned under software control into four equivalent subsystems, each of which consists of 16K ($= 16,384$) processing elements which can be operated independently under the control of four corresponding host processors. This "space-sharing," as opposed to the now conventional "time-sharing," capability facilitates program development activities. Fabrication of the 64K processor CM prototype was completed by December 1985. The machine is air-cooled, occupies about 25 square feet of floor space, and consumes about 20 kilowatts of power. A smaller subsystem is currently being used for software and applications development.

The CM hardware architecture is primarily intended to increase the effective utilization of the physical resources that are used to build a modern computer, in order to provide machines with a much greater level of performance than has hitherto been available. But hardware characteristics (such as the speed of execution of instructions and data transfers) constitute only one leg of the triad on which a successful computing system must be supported. The other two legs of the triad are the ease with which the computing system can be programmed, and the cost effectiveness of the system as a whole, including both the cost per MIPS executed and the cost of applications software development. Programming strategies for novel architectures emphasizing high concurrency of execution of instructions are very much terra incognita. Thus, it may be of interest to report some of the software concepts that underlie the design of the CM as a computing system rather than merely as a hardware architecture.

In general, the cost effectiveness of a computing system will be determined by how well the hardware and software architecture are adapted to the primary applications domains. The introduction of integrated circuit technology has greatly reduced the cost and design time required to build a new machine, and has consequently led to the development of special-purpose computers and attached processors optimized for one or a few applications. Floating-point and matrix-array processors, signal-processing computers, data-base machines, and VLSI circuit simulators are a few of the many types of machines that have appeared in recent years. Their cost effectiveness is testimony to the advantages of mating hardware and its associated software to the applications domain. This advantage of special-purpose machines is also their great limitation: they cannot be adapted to other tasks and so remain idle in the absence of a steady flow of work of just the right specialized nature. Thus, an interesting and potentially effective alternative strategy would be to design a machine whose functional structure could be dynamically reconfigured under *software* control in order to adapt it to the structure of the application. The CM hardware architecture supports dynamic reconfiguration under software control. Here, the issue of cost effectiveness reduces to a comparison of the cost of the excess execution time required to support the dynamic reconfiguration of the system, as opposed to the cost of supporting one or more special-purpose attached processors that are idle for a substantial fraction of the time. The balance will differ for different applications and computing environments, but the flexibility of the reconfigurable general-purpose computer is likely to be the dominant factor for researchers who are exploring the very paradigms of concurrent computation.

The CM can be dynamically reconfigured during the course of a computation. This is accomplished by identifying processing elements with natural elements of the application (e.g., *processor* —▸*picture element* for image processing; *processor*—▸*transistor* for circuit simulation; *processor*—▸*word* for language analysis), and by establishing a virtual software-controlled communication link between two processors if the associated problem entities would be connected in some graph-theoretic representation of the problem. Thus, processors representing the nearest neighbor picture elements in an image for computations such as edge detection, thresholding, etc. would be connected by a virtual communication link; two processors representing the transistors in a circuit would be connected by a link if the transistors they represent are connected by a wire; and processors representing words in a sentence might be linked to processors representing syntactic or semantic categories associated with the word.

The allocation of processors and the establishment of virtual links is performed automatically in the sense that the programmer need not be aware of the details of how the graph structure appropriate to the application is physically represented in the CM; the programmer deals with the relationship among graph vertices and graph edges in the particular applications domain at the symbolic level.

The CM system presently supports an extension of Common LISP called CM-LISP for high-level programming. In addition to the usual data structures and operations of LISP, CM-LISP contains several new operations expressly intended to support concurrent operations. These operations may be thought of as a generalization of vector algebra, where the vector operations correspond to operations on the components of one or more vectors that are mutually independent, and hence can be executed concurrently. If, in place of vectors, one considers functions whose domain and range are finite sets, then an ordering of the domain enables one to represent the effect of the function by listing its values in sequence. Operations including composition of functions, restriction of a function to a subset, and projection onto a subspace are conveniently incorporated into the general structure of LISP in a natural way.

The preceding description indicates that the connectionist viewpoint in the study of cognition is influencing the design of new parallel supercomputers. In their turn, powerful new machines that are particularly effective in supporting reseach in the cognitive and brain sciences may play a special role in advancing our understanding of the processes of thought and the nature of intelligence. The "fearful symmetry" of research into the nature of biological information-processing systems and electronic information-processing systems will, we may hope, reveal the general principles that are their common foundation, and thereby confirm Freeman Dyson's observation that " . . . as we understand more about biology, we shall find the distinction between electronic and biological technology becoming increasingly blurred."

A "fat tree" is a routing network for parallel computation whose underlying graph theoretical structure is a complete binary tree of vertices which has increasing communication capacity as one approaches the root of the tree. The leaf nodes of a fat tree represent processors and the other vertices of the graph represent communication switches, but for our purposes, there is no fundamental distinction between them. "Universal" fat trees can simulate any routing network having the same number of processors and requiring the same physical volume in linear time up to a power of the logarithm of the number of processors (Leiserson 1985). A universal fat tree is characterized by two parameters: the number n of processors and the root capacity, w. The number of vertices in the graph is $N = 2n - 1$. In a universal fat tree, the root channel capacity w is constrained by the inequality $n^{2/3} \leq w < n$. If we assume that w is of the form $w = n^a$, then the number of connections in the fat tree grows less rapidly than the Erdös and Rényi expression unless the parameter a is equal to approximately 0.89. Indeed, if $a = 2/3$, then the growth in the number of connections is only of order n. If a is greater than the critical value, then the corresponding fat tree has a high probability of remaining connected if some of its vertices or edges are randomly deleted.

Consider the popular (hyper) rectangular lattice architectures. Without loss of generality, we may restrict our attention to (hyper) cubical lattices. An array of N vertices arranged in a cubical lattice of side s in k-dimensional space satisfies the relation $N = s^k$, whence $k = \log N/\log s$. For convenience of presentation, let us suppose that corresponding vertices on opposite hyperfaces of the array are connected by edges so that the whole has the topology of a k-dimensional torus.[3] Then, the number of connections in the array will be

$$kN = \frac{N}{\log s} \log N$$

Suppose that N is large. The probability of the network remaining connected when some of its connections are eliminated will be great or small according as $1/\log s$ is greater than $1/2$ or smaller than $1/2$. Since $\log s = 2$ implies $s = e^2 = 7.38$, we see that the cubical lattice network will have a high probability of remaining connected after the failure of some of its parts only if $s \leq 7$; hence, the requirement of remaining connected requires that the lattice have a high dimension

k. This argues against the use of computer architectures such as 2-dimensional or 3-dimensional grid arrays for machines with very large numbers of processors.

Most robust among the cubical lattice networks are the boolean cube architectures. The vertices of this network correspond to sequences of k binary digits. Two vertices are connected in the network if and only if the corresponding binary sequences differ in exactly 1 bit. It follows that there are $N = 2^k$ vertices, $C = k2^{k-1}$ connections, and the ratio of connections to vertices is $k/2$. Each vertex is connected to k nearest neighbors

$$C = \frac{N}{2} \log_2 N = \frac{N}{2 \log 2} \log N \approx 1.4(\frac{N}{2} \log N)$$

This network has a high probability of remaining connected even if many of its parts fail.

If $k/2 = 16$, then the boolean 32-cube connects a system of $2^{32} \simeq 4 \times 10^9$ nodes, less than the number of neurons in the brain by a few orders of magnitude. Hence, the brain most likely has a lesser degree of connectivity than the boolean 32-cube.

Connectionist Models of Memory and Learning

The preceding discussion leads to a model of the brain as a tangled graph consisting of a large number of vertices—perhaps as many as 10^{12}—irregularly interconnected by perhaps 10 to 20 times as many edges. The vertices can be thought of as McCulloch-Pitts neurons (McCulloch-Pitts 1943), each capable of storing 1 bit of information which defines their state.[4] A connecting edge is capable of various degrees of excitation or inhibition of the vertex, or neuron, upon which it terminates. We thus come to view the role of the connections in the neural network, represented by the topology of the network and the possibly dynamically varying strengths of the individual excitation/inhibition interactions, as playing the dominant role in the processes of thought, with the vertices—neurons themselves—merely storing and forwarding state information.

At this point, one must stress the significance of the irregularities inherent in the patterns of connections of biological information-processing systems. Although the obvious gross anatomical similarities among structures serving similar functions in one organism and the same functions in different organisms usually capture our attention, the details of the "wiring diagram" fluctuate substantially. The manufacturing process lacks the vigorous quality control that would be required to assure that pairs of vertices are connected if and only if the "blueprint" calls for it. Nature has had to make a virtue of necessity by building in means for processing information that feature the random element in the interconnection topology. Thus, the mechanisms of brain operation are to a considerable extent independent of certain interconnection details: we observe complex behavior emerging from the unpredictable combination of grossly similar simple parts. The collective properties of interacting neurons constitute a kind of "statistical neurodynamics" to which the classical statistical thermodynamics of Boltzmann—which is nothing more than a theory of information in disguise—is broadly applicable. We may be reminded in this context of Schrödinger's characterization of statistical thermodynamics as the study of systems composed of large numbers of weakly interacting homologous units (Schrödinger 1964).

We take the weakly interacting "units" to be the vertices of the neural network considered as a graph. Each neuron may be supposed to exist in one of two possible states. The state of the ith neuron will be denoted by S_i where $S_i = 0$ if the neuron is off, and $S_i = 1$ if it is firing at its maximum rate. The state of the neural network is represented by the binary vector $S = S_i$. The interactions between pairs of neurons are described by the strength of the excitation or inhibition between vertices linked by an edge. This is the point of view that has been explicitly adopted by Hopfield in his development of a connectionist theory of memory and by Hinton and his colleagues in their pursuit of a theory of learning. It is also implicit in "spreading activation" theories, such as that of the Waltz and Pollack (1985) model of language understanding (see also Feldman 1982; Feldman and Ballard 1982; Rosenblatt 1961).

Thus, statistical thermodynamics, in the formulation endowed upon it by Boltzmann, appears as the natural theoretical structure for thinking about thinking.

Let us pursue this train of ideas, following Hopfield (1982) and Ackley et al. (1985).

The "strength" or "weight" of the connection between neuron i and neuron j will be denoted by W_{ij}. If the ith and j^{th} neurons are not connected by an edge, then $W_{ij} = 0$ for all i. If t_i is the threshold for neuron i, then we may suppose that i changes state according to the prescription

$$S_i \rightarrow 1 \text{ if } \sum_j W_{ij}S_j > t_i$$

$$S_i \rightarrow 0 \text{ if } \sum_j W_{ij}S_j \leq t_i$$

Changes of state occur randomly and asynchronously, but at a fixed average rate.

In order to bring the full power of the machinery of statistical thermodynamics to bear, we must introduce a generalized cost or "energy" function, and also the information or "entropy" associated with the system.

The energy of a configuration of the network (i.e., corresponding to the system of states $(S_i, i = 1, \ldots, n)$ is a quadratic form in the elements of the vector S: $E = -S'WS$. Here, $S = S_i$, $i = 0,1, \ldots, n$ where $S_0 = 1$ and $W = (W_{ij})$, $0 \leq i, j \leq n$, where $-2W_{i0} = -2W_{0i} = t_i$ is the threshold for neuron i.

Such a system can "remember" configurations, and it can also associate with an arbitrary new configuration one of its memories which approximates the new configuration: it can serve as a *content addressable memory*. For instance, let $S^m:m = 1, \ldots, k$ be configurations to be stored as memories. Then, selection of the weights according to the formula

$$W_{ij} = \sum_{m=1}^{k}(2S_i^m-1)(2S_j^m-1), i \neq j$$

defines an energy function E for which the states S^m are local minima. Hopfield finds that about $0.15n$ configurations can be remembered before recall becomes severely compromised.

The values of the components of S can be taken to represent the acceptance or rejection of some hypothesis about a domain under consideration: $S_i = 1$ may represent acceptance of the hypothesis and $S_i = 0$ its rejection. Suppose that W is a symmetric matrix. If i and j are linked with weight W_{ij}, then the hypotheses i and j are mutually supportive if $W_{ij} > 0$ and mutually antisupportive (i.e., acceptance of one hypothesis supports rejection of the other) if $W_{ij} < 0$. If $W_{ij} = 0$, the hypotheses are independent. With this interpretation, the energy of a configuration represents a measure of the aggregated support for the hypotheses.

If we think of the multidimensional surface defined by the energy as a function of the state vector elements, then the "memories" S^m correspond to local minima—to valleys in the surface—which can be reached by standard gradient descent methods. If an input state vector lies close to some local minimum, gradient descent will find that minimum, and thereby identify the input state with it. In this sense, the network can be said to be capable of both storing new memories (by changing the energy function to include them) and of identifying a new impression with a stored memory, although the two may be far from identical. The model may thereby be capable of accounting for the puzzling ability of the brain to classify percepts that are dissimilar in their details into classes that have some important gross similarity.

   Application of network models to evaluate hypotheses leads to a rather more complex and subtle situation, for, in this case, what is desired is not a local minimum of the energy function—which would represent acceptance of the given hypothesis relative to nearby states of the configuration—but rather a global determination of whether the hypothesis should be accepted that takes into account the totality of support and rejection, direct and indirect, that is represented by the combination of topology and weights of the connections. It is well known that local methods of analysis, such as gradient descent, are inadequate for this purpose. It is at this point that the other fundamental concept of statistical thermodynamics enters: the *information*, or *entropy* function. The time evolution of thermodynamical systems carries them forward toward a stable state that is characterized by maximum entropy. An equivalent and more intuitive characterization is that the measure of information associated with the state of the system assumes its minimum value in an equilibrium state. The system fluctuates about this minimum due to random variations in the values of the state variables. If the possible states of the system are indexed by Greek subscripts, and if P_α denotes the probability of occurrence of state S_α, then the measure of information associated with the collection of possible states is

$$I = - \sum_\alpha P_\alpha \log_2 P_\alpha$$

In the equilibrium configuration of minimum information, the probabilities are related to the energy by the Boltzmann distribution formula,

$$\frac{P_\alpha}{P_\beta} = \exp\left(-\frac{(E_\alpha - E_\beta)}{T}\right)$$

where E_α is the energy corresponding to the state S_α and T denotes a parameter which is determined by the structure of the system and plays a role analogous to temperature in thermodynamics. When T is large, the relative probability of finding the system in a state of relatively high energy increases, as is indicated by the formula; if T is close to zero, then the system spends most of its time in the lowest available energy state. This, of course, corresponds precisely to the physically analogous situation, in which it is possible to find a thermodynamical system in a state of high energy if the temperature is great enough, because then the energy transfer between the homologous parts is correspondingly increased.

   This is not the place to discuss the still poorly understood relationship between entropy and information, which has been the subject of many profound insights and even more numerous confused mis-interpretations. For the present purpose, it will suffice to note that the algorithms for calculating a state that corresponds to the global min-imum of the information function constitute a numerical simulacrum of the process of physical annealing: just as a physical system at high temperature can be made to jump from one to another "valley" on the energy surface and will, as the temperature is decreased, settle into a configuration whose energy is, or is close to, the global minimum as it "crystallizes," so too can the neural network model be made to "crystallize" into the state that corresponds to the global acceptance of a hypothesis as the consequence of a large number of small random interactions throughout the network. The reader is referred to the literature for the details (cf. Ackley, Hinton, and Sejnowsky 1985). Here, it suffices to note that, just as in statistical thermodynamics, the cal-culations that determine the desired state are individually simple, al-though large in number, and are arranged so as to play themselves out in parallel as the inputs and other environmental conditions in which the system finds itself dynamically change.

The Connection Machine™ Computer

A science progresses most rapidly when it has experimental tools that enable the researcher to observe and measure phenomena with a refinement that exceeds the capabilities of the unaided senses. This is perhaps one reason why the study of the processes of reason and the nature of intelligence has progressed so slowly in the past. The com-puting machine has offered for the first time the opportunity to test the predictions of models of cognitive phenomena. But in our current state of knowledge, the phenomena appear to be so complicated that even the most trivial models of them tax the power of the largest machines that are available today. In particular, it appears that the physical limitations of microelectronic technology will make it impos-sible to develop computing machines of conventional serial architecture that possess sufficient power for the study of these problems. Thus, the cognitive scientist should welcome the recent efforts of the computer architect to develop designs for parallel computers that might be par-ticularly well suited to the investigation of cognitive processes. We will describe one parallel computer architecture that has its intellectual roots in the connectionist point of view and was designed with the express objective of providing a tool well matched to the needs of the cognitive scientist.

A computing machine consists of three essentially distinct parts: a *memory* unit which is used to store data, the program, and intermediate results; a *processor* which performs the arithmetical and logical operations that are specified by the program; and a *communication network* that connects the memory and processor components of the computer. This view is an oversimplification: the three subsystems are neither conceptually nor physically entirely distinct from one another, and the complete system includes input and output channels which provide means for extending memory and modifying the data and program.

In early realizations of computing machines, the processor, memory, and communication components were fabricated from different materials: vacuum tubes, magnetic drums, and wires provided the technology base for machines built 30 years ago. One important consequence of the development of large-scale, semiconductor integrated circuit technology has been to eliminate the physical distinctions in the materials from which the processing, memory, and communication elements are fabricated (Mead and Conway 1980). Each of these key constituents is composed as a layered pattern of conducting and insulating regions on the surface of a thin wafer of silicon or other semiconducting material. Thus, the physical properties of the three types of components are reduced to the geometrical properties of the circuit patterns that correspond to them on the surface of the semiconductor material. In particular, the cost of a particular component is, in effect, measured by the area it occupies on the chip. This reduction of microelectronics to geometry has led to the development of a mathematical theory of integrated circuit design that allows the circuit engineer to study optimal designs and to prove theorems about them. It also provides a valuable tool for thinking about computer architectures.

From this point of view, the cost of manufacturing a computer is, apart from the input/output and peripheral devices associated with it, measured by the cost of fabricating the integrated circuits of which it is composed, and that cost, in turn, is, for a given processing technology, proportional to the area of the silicon or other semiconductor used. The problem faced by the computer architect is to obtain the greatest level of performance (assuming a given processing technology) from a fixed area of silicon by the efficient arrangement of the patterned layers into processing, memory, and communication elements.

Consider a traditional serial—a so-called von Neumann—computer. The integrated circuits in a contemporary "mainframe" computer may have an aggregated area of about 1 square meter of silicon, which incorporates the processing elements, the main high-speed memory, and the memory decoder communication network used to collect program instructions and data from memory and transmit the results produced by the processor to the memory.

The processor is a single, localized, powerful unit which can at any time execute one of a large number of instructions. It typically occupies about 5% of the silicon area. Memory occupies more than 70% of the area, with the communication network consuming the remainder.

The typical duty cycle of the serial computer consists of the repetition of a standardized sequence of events: an instruction is *fetched* from a location in memory; a pair of operands is *fetched* from memory; the result of applying the instruction to its operands is *stored* in a memory location; and the cycle is repeated. Thus, the communication network is always busy decoding memory location references and storing and fetching information, with that part of the processing unit active every duty cycle, but most of the memory is *inactive*: there are only four memory accesses per duty cycle for a typical instruction. Only a small fraction of the silicon is active during the course of computation. Considered from an abstract point of view, the space-time volume of silicon devoted to communication is relatively active, and hence, relatively efficiently utilized; the space-time volume devoted to the processor is less effectively utilized because only one instruction circuit is active at a time from among the many hundreds of instructions that the processor is capable of executing; and the space-time volume corresponding to the memory is largely inert.

The trend toward employing computers to aid in the analysis of ever more complex problems involving large amounts of data has led to a systematic increase in the quantity of high-speed memory incorporated in the machine. This further reduces efficiency, because the fraction of the silicon area that is active at a given instant decreases as the size of the memory increases.

Notes

1. The author is indebted to W.D. Hillis for bringing the theorem of Erdös and Rényi to his attention.

2. We will systematically ignore the negligible difference between C_a and the expression obtained from its definition by omitting the square brackets denoting "the integer part of."

3. The number of additional connections grows more slowly with increasing n than does the number of connections in the cubical lattice, and therefore does not influence the connectivity of the array.

4. In the absence of definitive information, we assume that the state of a neuron is described by one bit, although more than one may actually be required. In the latter event, the graph would be replaced by one with a larger number of vertices, each of which would represent a state defined by one bit.

Bibliography

Ackley, David H., Geoffrey E. Hinton, and Terrence J. Sejnowsky, 1985: A learning algorithm for Boltzmann machines, *Cognitive Science* 9:147–169.

Broomell, George, and J. Robert Heath, 1983: Classification categories and historical development of switching topologies, *ACM Computing Surveys* 15:95–133.

Erdös, P., and A. Rényi, 1959: On random graphs I, *Publ. Math. Debrecen* 6:290–297.

Feldman, J.A., 1982: Dynamic connections in neural networks, *Biological Cybernetics* 46:27–39.

Feldman, J.A., and D.H. Ballard, 1982: Connectionist models and their properties, *Cognitive Science* 6:205–254.

Haynes, Leonard S., Richard L. Lau, Daniel P. Sieworski, and David W. Mizell, 1982: A survey of highly parallel computing, *IEEE Computer,* January, 9–24.

Hillis, W.D., 1985: *The Connection Machine,* MIT Press, Cambridge, Mass.

Hopfield, J.J., 1982: Neural networks and physical systems with emergent collective computational abilities, *Proceedings of the National Academy of Sciences USA* 19:2554–2558.

Leiserson, Charles E., 1985: Fat-trees, *Proceedings IEEE International Conference on Parallel Processing.*

Mead, Carver, and Lynn Conway, 1980: *Introduction to VLSI Systems,* Addison-Wesley, Boston.

McCulloch, Warren, and Walter H. Pitts, 1943: A logical calculus of the ideas imminent in nervous activity, *Bulletin of Mathematical Biophysics* 5:115–133.

Rosenblatt, F., 1961: *Principles of Neurodynamics—Perceptrons and the Theory of Brain Mechanisms,* Spartan, Washington, D.C.

Schrödinger, E., 1964: *Statistical Thermodynamics,* Cambridge University Press, England.

Waltz, D.L., and J.B. Pollack, 1985: Massively parallel parsing—strongly interactive model of natural language interpretation, *Cognitive Science* 9:51–74.

TABLE 6.1 Supportable Connectivity of $n = 10^{12}$ Vertex Graphs
for Various Switching Energies

Technology	Feature Size	E_{sw}	C	C/n
1978 CMOS	6μm	6.24×10^6 ev	5.1×10^9	5×10^{-3}
1985 CMOS	2μm	2.3×10^5 ev	1.4×10^{11}	1.4×10^{-1}
19XX CMOS	1μm	2.9×10^4 ev	1.1×10^{12}	1
Limits of Physical CMOS Technology	0.3μm	1.25×10^3 ev	2.6×10^{13}	26
Thermal Background at 300 Kelvins	--	$kT = 0.026$ ev	1.3×10^{18}	1.3×10^6
Single Channel Neural Current	--	2×10^3 ev	1.6×10^{13}	16

TABLE 6.2 Variation of p(a)

a	p(a)
-2	1.9×10^{-24}
-1	0.0006
-0.5	0.0659
0	0.3678... $= 1/e$
1	0.8734
2	0.9818
3	0.9975
4	0.9996
5	0.9999

7. Evolutionary Learning of Complex Modes of Information Processing

Introduction

In Hastings and Waner (1985), the authors developed principles of evolutionary learning and described a model for a stochastic neural network which realized these principles. Briefly, an evolutionary learning system in this sense incorporates a state space whose individual states are modes of information processing, and an evolutionary search through these states which follows annealing dynamics.

The neural network model therein described was capable of realizing the evolutionary learning of arbitrary directed graphs, and thus sufficed to model the learning of stimulus response behavior. The net was structured as a 2-dimensional lattice, with each node connected to others in a suitably defined neighborhood. The nodes themselves were McCulloch-Pitts neurons with a firing threshold of 1. The edges modelled synapses along which conduction was stochastic and mediated by conduction thresholds. (Only excitory stimuli were used.)

The net lacked sufficient structure to learn complex logical processing modes which incorporate such primitives as "and" and "not." The first objective of this chapter is to enrich the net of Hastings and Waner (1985) so as to permit the evolutionary learning of complex processing modes. This will be achieved by including a form of "bit masking" as a mediating structure. The existence of such masking mechanism in the brain will be discussed later.

The introduction of bit masking will be seen to appropriately (and exponentially) enlarge the state space. This has the concomitant effect of exponentially increasing the complexity of ergodic searches. Thus, a second objective of this chapter is to propose two natural speedup mechanisms in order to achieve a satisfactory learning rate despite this apparent difficulty. Without such speedup mechanisms, the learning time would increase exponentially with problem size. Such experimental increases would rapidly render intractable the solution of large problems.

A Stochastic "Generation 1" Neural Network

As already mentioned, our neural network will incorporate the model in Hastings and Waner (1985), together with a superimposed bit-masking structure at each node which mediates the activity of that node. Here, we recall facts about the underlying "generation 1" neural network and design philosophy.

Hastings and Waner (1985) describe an evolutionary learning system as an ergodic system whose states are modes of information processing. Such a system was based on three "principles of evolutionary learning." The first principle is essentially the statement that the system searches ergodically through the state space, thus passing arbitrarily close, given sufficient time, to any particular state. The second principle is that of annealing, and will be discussed at some length later. Briefly, the principle asserts that the trajectories of the system tend to be driven toward states which correspond to minima of a potential function defined on the state space, and that the degree of stochastic behavior is mediated by a parameter analogous to temperature, which tends to lower as the system adapts. The third principle is the assertion that the potential energy corresponds inversely to the extent of adaptive (or desired) behavior, as determined by interaction with the environment. Such a learning dynamic is termed "soft programming," as it is behaviors—rather than the explicit modes of information processing giving rise to them—which are re-enforced by this strategy. (This is also discussed further later.) The dynamics of such a system drive it locally toward deterministic behavior modes as the temperature lowers, and preferred behavior modes are selected if the local "cooling" toward determinism is sufficiently slow to ensure that the system is not trapped in a local minimum.

Our implementation takes the form of a dynamic model whose degree of (local) determinism is governed by state parameters. The configuration space of these parameters is convex, with extreme points corresponding to the possible deterministic behavior modes. The dynamics drive the system toward extreme points of its parameter space locally ("learning/cooling"), and toward the interior elsewhere ("forgetting/reheating").

The net takes the form of a rectangular lattice in which each node represents a formal neuron with activity 0 or 1, and a firing threshold of 1. Each neuron is connected to others in a neighborhood (for example, to first- and second-nearest neighbors) by formal synapses possessing variable and independent conduction thresholds.

If the activity of a neuron is 1, the neuron fires to each of its neighbors with a probability given by prob(conduction) = 1 − (conduction threshold), transmitting an activity of 1 to each target neuron and losing its own activity in the process. Firings take place simultaneously (in some variants of the model), and are regulated by a clock. When a synapse conducts, its conduction threshold is decreased by a (variable) amount, Δ, making it more likely to conduct, while synapses which fail to conduct have their conduction thresholds raised. Superimposed on this is a decay mechanism in which thresholds decay with time to a predefined base value.

Simple structures such as directed paths and graphs may now be learned by the feedback mechanism described in Hastings and Waner (1985). Briefly, selected source nodes are activated, and a response is awaited at selected target nodes. If such a response fails to occur after a preset waiting period, the sources are reactivated. When the desired response occurs, the sources are then immediately reactivated (before appreciable threshold decay has occurred), and the process repeated until the desired path or graph has "burned in."

Since the deterministic modes in this net are directed graphs without additional structure, the structure so far precludes the development of "and" and "not" primitives, and this we now remedy.

Deterministic and Stochastic Bit Masking

The deterministic modes in our system will include the additional structure of deterministic bit masks. Stochastic bit masks will be defined using suitable convex combinations of deterministic bit masks, and learning will involve driving the system (locally) toward deterministic modes.

First, we discuss bit masking in a deterministic setting. Thus, one has a net in which all edge thresholds are either 0 or 1. Fix a neuron n, and let N(n) be the set N(n) = (neighbors of a n connected to n via a single incoming edge). *A deterministic bit mask* is then a map $m:N(n) \rightarrow [0,1]$. This defines a vector m indexed by the elements of N(n). At any instant t, the incoming signals define a similarly indexed stimulating vector, s. Let a(s,m) be the number of elements in N(n) on which m and s agree. Also assume that one has associated with each neuron n a firing threshold, T(n). If, at that instant, a(s,m) $\geq$ T(n), then the node fires.

One can now realize "and," "not," as well as more complex gates within a net including this structure by choosing suitable values for s, m, and the vector γ of incoming thresholds. To demonstrate this, we list a few examples.

Examples

1. At a site n, let m be the vector with each coordinate set to 1, and let T(n) = 1. Then n acts as an "or" gate, firing upon receipt of any input configuration s with one or more non-zero coordinates. One may further select the specific inputs over which to "or" by setting the conduction thresholds associated with those inputs (incoming edges) to 1, and those associated with the remaining inputs to zero.

2. A "not" gate (with a single input variable) may be realized by masking the (only) non-zero incoming edge with 0 and setting T(n) = 1 + (number of zero incoming edges). By including more than one non-zero incoming edge, one can "or" over any number of "nots." For example, with two non-zero incoming edges each masked with 0 and T(n), one has "γa or γb."

3. To achieve "and" gates, one increases the firing threshold, T(n). Thus, one may realize $(a_1 \, \& \, a_2 \ldots \& \, a_k)$, with k non-zero incoming edges masked with 1 and a firing threshold of k + (number of zero incoming edges).

4. By maintaining the firing threshold as in Example 3 and altering the bit mask, one can induce a strict masking. Thus, a neuron will fire only if the stimulating configuration agrees with that of the mask.

5. More complex behaviors, such as "two-out-of-three" gating, are achieved by using firing thresholds between 0 and k + (number of zero incoming edges).

6. "Trivial" gates are obtained by setting T(n) either to 0 or to (number of incoming edges) + the former resulting in a gate that always fires, and the latter in one that never fires.

Thus, given sufficiently many nodes and connections, arbitrary logical circuits may be realized. We now consider the stochastic case. A stochastic bit mask will be thought of as a probability measure on the set of deterministic bit masks. For reasons to emerge later, the probability measure associated with a stochastic bit mask will be defined somewhat less directly.

Let n be a neuron, and let N(n) be its neighborhood. A *stochastic bit mask* is a map $r{:}N(n) \longrightarrow [0,1]$, where [0,1] is the unit interval. A stochastic bit mask r determines a probability measure $\mu(r)$ on the set S of deterministic bit masks by defining

$$\mu(r)(m) = \prod_x \{(1-m(x))(1-r(x))+m(x)r(x)\}$$

where product is taken over all elements x in N(n), for a given deterministic mask m. This says that, at any instant, the bit mask defined by r has the value 0 at x with a probability $(1 - r(x))$ and the value 1 with probability $r(x)$, the behavior at the element x being independent.

In the same spirit, we arrange for the firing threshold at a given neuron to be a probability measure F, on the collection of all possible thresholds, which, in this case, we define directly. Thus, let N(n) have k elements. The measure F is then given by a tuple $(p_1, p_2 \ldots, p_k)$ with $0 \le p_i \le 1$ and $\Sigma_i p_i = 1$. The nodes then behaves as a node with firing threshold i with probability p_i.

Note that the stochastic structure becomes deterministic at the extreme points of its parameter configuration space, extending the analogous generation 1 structure. In the next section, we discuss the dynamics of the model. We shall refer to a net incorporating this structure as a "generation 2" net. Structurally, a generation 1 net incorporates simple learning (in our case, of connected graphs), while a generation 2 net incorporates behavior modes with arbitrary logic.

The Learning Dynamic

Here, we describe modification and decay mechanisms which drive the learning dynamic in a net incorporating the features discussed previously. The basic philosophy here agrees with that of the underlying first generation net; a topologically interior point of the system configuration space is chosen as an initial value, and the dynamics are such that extreme points play the role of attractors locally.

With this philosophy in mind, the initial net parameters may be chosen as follows. For the values of each $r(x)$, with x an incoming edge, choose a constant value between 0 and 1 as a base value. As time progresses, value decay to this base along an S-shaped decay curve. If a neuron n fires in response to a stimulus vector s, the value of $r(x)$ is raised or lowered toward $s(x)$, (which, we recall, is 0 or 1), by an amount Δ proportional to $(1 - $ distance from $s(x))$. Thus, the value of each $r(x)$ is driven toward an extreme value (0 or 1 in this case). (See Figure 7.1.) Further, the proportionality factor effectively places the extreme points of the system infinitely far from any interior point, thus ensuring that the system never becomes deterministic. This is analogous to the use of the Shashahani metric in genetics.

The probability measure F, defined on the set of possible firing thresholds, must be made to behave similarly. One begins with an interior value—say, all probabilities equal—and then skews the measure toward one of the characteristic measures in response to firing upon the corresponding value of a(m,s), where m and s are as earlier described. Similarly, decay tends to drive the threshold measure back toward the base.

In order to motivate this mechanism, we consider, as an example, a 4-neuron segment (of a generation 2 net), which can learn "and," "or," or "not."

Example

Learning an "and" gate. We consider a 4-neuron segment of the net, with connections represented as shown in Figure 7.2. (Other interconnections, and connections to the rest of the net, are not shown, but will be assumed present.) Beginning with edges 1 and 2 masked with values of 1/2, one stimulates B and C simultaneously, and awaits a response at A'. If no response occurs for a time T comparable to the decay-time corresponding to a single-step modification, the event may be thought of as "forgotten," and the sources are again stimulated. Since there is a non-zero probability of a response at A' within a time significantly shorter than T, there will be an eventual response at A', presumably from A. At this instant, A has its mask temporarily skewed toward the vector (1,1), and its firing threshold skewed toward 2. (We are at the moment ignoring external interference for the sake of simplicity.) Thus, restimulating B and C immediately increases the likelihood of a repetition of the same behavior. This being done, the process is continued until the response at A' is reliable. Since the firing threshold at A is now reliably close to 2, the gate will seldom function as an "or" gate, unless such behavior is re-enforced by the same restimulation regimen.

Since this system is not an isolated one, account must also be taken of the net environment. (For example, the signal need not have travelled via A.) One can nonetheless argue that the eventual, almost deterministic, local behavior will be that of an "and" gate for two reasons. First, whatever local behavior has resulted in the response will tend to repeat as the system parameters drift locally toward their extreme values. Second, the combination of bit-masking and firing threshold parameters diminish the likelihood of the same response occurring as the result of a different stimulation pattern.

Learning an "or" gate. Here, one begins to alternately stimulate A, B, or both upon a response at A, without a long-term preference for any one stimulation pattern. This causes the bit masks (0,1), (1,1), and (1,0) to become preferred, but not, however, equally likely, even though the three corresponding stimulation vectors are presented with equal frequency. This apparently anomalous behavior results from the way in which the bit-mask probability measure is defined: the coordinates 1 and 2 of the stochastic bit-mask vector both approach 1, resulting in an (ultimately) deterministic preference for mask (1,1). The corresponding firing threshold is seen to be 1 two-thirds of the time. Thus, one has an ultimately deterministic "or" gate. Note that the particular way in which stochastic bit masks were realized plays an essential role in the possible evolution of "or" gates. Had the stochastic bit mask taken the form of a simple probability measure on the set of deterministic masks, with skewing taking place toward the characteristic measure of the predominant stimulus s, no "or" gate could evolve by this dynamic, since the vector s is varying throughout the learning process.

Learning a "not" gate. Here, one may stimulate B only, and reward upon response at A'. The deterministic parameters then include a (1,0) gate at A and a firing threshold of 2. Thus, if both B and C are stimulated, the node A will not fire. The role of the signal from B may be thought of as an activation signal, so that the situation represents a "fire on not C in response to activation B" behavior. (In our model, a neuron will never fire unless activated by at least one neighbor.)

While the learning of relatively simple behaviors such as this pose none of the exponential growth problems associated with large random searching, it is easy to see that the learning of more involved tasks brings such difficulties to the fore. Since this problem has apparently been solved in the dynamics of evolutionary genetics, we consider in the next sections ways in which to exploit this in our net.

Annealing

In Hastings and Waner (1985), we observed that evolutionary learning is an annealing process. The method of computational annealing used in solving certain optimization problems can be viewed as intermediate between gradient methods and random search methods. Moreover, annealing dynamics can be interpolated between the dynamics of both extremes (gradient and random search) by adjusting either the potential energy surface or the parameter representing temperature.

More formally, consider the problem of minimizing a potential function V defined on a state space S. Assume for now that S is a smooth manifold and that V is a smooth function. One may then locate a local minimum of V by the following procedure. Given a present state s, compute the gradient, grad V(s). If grad V(s) = 0, then s is a critical point. All local minima are critical points; in practice, one can determine whether a given critical point is a local minimum. If s is a local minimum, we are done; otherwise, choose another state s' as starting point. In the case grad V(s) ≠ 0 (which is most likely for a random starting point), one then moves to a new state s' given by s' = s − gradV(s) ε for a suitably small ε. This process specifies what we call a "gradient automaton." Gradient automatons typically solve local minimization problems in polynomial time. Thus, in the event that all local minima in V are also global minima, a gradient automaton may be considered as an efficient device to be used in locating global minima. However, if there are several local minima which are not global, the gradient automaton fails in general to locate global minima.

This setting may be significantly generalized to include discrete minimization problems, including linear and convex programming. In the context of linear programming, what we have described is essentially the simplex method, which runs in average polynomial time, although some cases require exponential time. (Other methods show that linear programming problems can always be solved in polynomial time.)

A second extreme method is that of random search. This method is most easily described for satisfiability problems. (Note that satisfiability and optimization problems have equivalent computational complexity, cf. Garey and Johnson 1979.) We pose our satisfiability problem as follows: given S and V as earlier described and given a bound B, find a point s such that V(s) ≤ B. A random search then proceeds as follows. Given a current state s, if V(S) ≤ B, the problem is solved. Otherwise, choose a new state s at random, compute V(s), and repeat. We shall refer to an automaton which performs such searches as a "random search automaton." Although random searching may strike one as inefficient, there are many interesting problems (satisfiability, traveling salesperson, integer programming, various games) for which no essentially better methods are know. These problems fall in the classes NP and PSPACE (cf. Garey and Johnson 1979). Each of these classes contains "universal" problems, called respectively, NP-complete and PSPACE-complete. No polynomial time solutions are known for such problems, and the outstanding conjecture that P ≠ NP (and thus, P ≠ PSPACE, a larger class than NP), asserts that no polynomial time solutions exist. If, as seems likely, the conjecture is true, then universal deterministic algorithms to solve such problems (on standard computers) would be computationally intractable due to a combinatorial explosion, and essentially no better than random searching. However, one should note that specific problems of this type may be combinatorially tractable in view of the fact that the expected waiting time t is proportional to 1/f, where f is the fraction of satisfactory states as determined by the measure used in the random search. Maynard-Smith (1970) and Conrad (1979) have explored the difficulty of evolution in this context, using Smith's "protein space" as the state space.

Note that the effectiveness of gradient searching rests upon the use of information about the present state. However, the use of purely local information can lead to the system halting at nonglobal minima. Random searching, on the other hand, avoids this problem by ignoring local information (although it is precisely this lack of useable information which renders random searching combinatorially intractable in general). Annealing combines some of the advantages of both methods. Following is a brief outline of the process of annealing in this context. (The reader is referred to Metropolis et al. (1953) for the original concept and to Kirkpatrick, Gelatt, and Vecchi (1983) for a survey and computational details.) Other references include Geman and Geman (1983) for applications to adaptive pattern recognition, Hinton, Sejnowski, and Ackley (1984) for an annealing machine which learns to solve satisfiability problems; Hastings and Waner (1984) for dissipation in annealing machines; and Hastings and Waner (1985) for annealing in the context of evolutionary learning.

Given a present state s, an annealing automaton chooses a new state s' among states s'' in a suitable neighborhood of s with probability depending on a comparison of $V(s')$ with the set $(V(s''):s'$ is a neighbor of s), and a parameter T thought of as temperature. In the limiting case of very large T, transitions to all neighboring states are equally likely, so that, in this extreme case, an annealing automaton is a random search automaton which searches neighboring states. As T approaches zero, the behavior of an annealing automaton approaches that of a gradient automaton. In general, the probability at temperature T of a state transition $s \longrightarrow s'$ involving an energy change $\Delta V = V(s') - V(s)$ is proportional to $_e - \Delta V/kT$, where k is Boltzmann's constant. Thus, the most interesting behavior occurs at intermediate values of T, where V is lowered with high probability. However, V may be increased (with low probability), thus permitting the system to avoid being trapped in local minima. Figure 7.1 illustrates this behavior. The annealing process thus combines the speed advantages of gradient flow with the ability, by means of random searching, to escape local minima.

The dynamics of annealing depend heavily on the shape of the potential contour: whereas annealing proceeds rapidly on many potential surfaces, it is effectively no better than a random search in the case of a potential function of the form

$$V(s) = \begin{cases} -1 \text{ if } |S-S_o| < \varepsilon \\ 0 \text{ otherwise} \end{cases}$$

where ε is small.

Many annealing contours allow gradient flow methods to locate minima rapidly, and for such contours, annealing works effectively when coupled with a suitable "cooling" schedule. Geman and Geman (1983) have described circumstances and cooling schedules under which annealing is computationally tractable. It has also been found empirically by many authors (Hinton et al. 1984; Hastings and Waner 1985) that annealing works well to solve general, as well as evolutionary, learning problems. Metropolis considered the problem of computing equations of state, Gelatt cites problems such as the travelling salesperson problem and the chip wiring problem, and Geman and Geman consider the problem of adaptive pattern recognition.

Speedup Techniques

The efficiency of annealing techniques may be increased by a suitable formulation of the problem to be solved. Here, we consider speedup methods applicable in general adaptive learning via annealing, and in our model.

In keeping with our philosophy of "soft programming" (Hastings and Waner 1985), desirable behavior modes are not necessarily unique states of the learning system, and one is essentially reformulating the search problem from one involving a search among states (information-processing modes) to one involving a search among behaviors. Since the mapping from states to behaviors is generally a "many-to-one" mapping, one is therefore effectively reducing the size of the state space being searched (cf. Conrad 1979 for evolutionary searching in the context of protein spaces). In soft programming, the programmer guides behavioral goals of the system, rather than its evolution through states. (Note that soft programming therefore allows the system to conduct many parallel searches for states which yield the desired behavior, without placing excessive demands on the input-output channels of the system.) Thus, the combinatorial explosion which occurs when complex modes are introduced is largely ameliorated by the corresponding combinatorial explosion in redundancy of the mapping from states to behaviors. Thus, soft programming may be viewed as a speedup obtained by, in effect, projecting the state space onto a smaller space.

Another speedup technique may be roughly described as one of "partial credit." In its simplest form, this amounts to the learning of a complex task through small and simpler steps. This kind of speedup was described in the context of silent genes by Hastings and Pekelney (1982). Specifically, assume that the expected waiting time for a single point mutation to occur is given by t. Hastings and Pekelney observed that a mutation which required N separate point mutations, each combination of fewer than N of which was deleterious, could occur in time $O(Nt)$ rather than $O(t^N)$—in effect, an exponential speedup if N is interpreted as the problem size. In the context of annealing, one may interpret this phenomenon as follows. Consider a state space S which is the product of N separate spaces

$$S = S_1 \times S_2 \times S_3 \times \ldots \times S_N.$$

Assume also that V is the sum of N separate potential functions, V_1, $V_2 \ldots V_N$, each defined on the corresponding state space, with

$$V(s_1, s_2 \ldots s_N) = \Sigma_i V_i(s_i).$$

Then, the problem of minimizing V reduces to the independent problems of minimizing the V_i. In the simple case that each of the N state spaces has the same size, this yields a speedup which is exponential in N. This situation approximates what occurs under a regimen of rewarding learning with partial credit for behaviors viewed as components of the desired behavior.

Further speedups are attainable by attempting to minimize all of the potential functions V_i in parallel. Such a speedup is especially useful in the context of biochemical information processing, where time is relatively valuable as a result of the slowness of the processors (neurons, groups of neurons, and even Darwinian reproduction) involved. The space, however, is relatively plentiful (the human brain has 10^{11} neurons and 10^{15} synapses).

Redundancy and distributed storage may be used to attain a further speeding up; previously learned functional components already may be stored as a result of redundancy and distributed storage. Thus, learning appropriate tasks may be viewed as minimizing V, given that one or more of the V_i are already minimized.

In short, soft programming, large redundancy, partial credit, previously learned components, and distributed storage—all of which are features of natural evolutionary processes—suggest powerful techniques exploitable in simulated annealing based on large state spaces.

Discussion

Evolutionary learning is appropriately viewed as an annealing process in which the states of the system are modes of information processing, with the states corresponding to preferred behaviors being assigned low potential energy. In order that such a system be capable of learning complex modes of information processing, these must be realizable within the learning system state space. Although inclusion of such modes leads, in principle, to a combinatorial explosion, it is proposed that the features associated with a suitable system—such as soft programmability and large-scale redundancy—together with appropriate learning regimens, may lead to combinatorial tractability.

In previous work, Hopfield (1982) studied the stochastic mechanics of large formal networks, and, in particular, the problem of information storage. Geman and Geman (1984) introduced the use of annealing machines to solve problems of image enhancement and pattern recognition. These ideas were further developed and formalized by Hinton, Sejnowski, and Ackley (1985) under a theory of "Boltzmann" machines.

Bibliography

Conrad, M., 1979: Bootstrapping on the adaptive landscape, *BioSystems* 11:167–182.

Garey, M.R., and D.S. Johnson, 1979: *Computers and Intractability: A Guide to the Theory of NP-Completeness*, W.H. Freeman, San Francisco.

Geman, S., and D. Geman, 1984: Stochastic relaxation, Gibbs distributions, and the Bayesian restoration of images, *IEEE Transactions on Pattern Analysis and Machine Intelligence* 6:721–741.

Hastings, H.M., and R. Pekelney, 1982: Stochastic information processing in biology, *BioSystems* 15:155–168.

Hastings, H.M., and S. Waner, 1984: Low dissipation computing in biological systems, *BioSystems* 17:241–244.

Hastings, H.M., and S. Waner, 1985: Principles of evolutionary learning—design for a neural network, *BioSystems* 18:105–109.

Hinton, G.F., T.J. Sejnowski, and D.H. Ackley, 1985: Boltzmann machines—constraint satisfaction networks that learn, *Cognitive Science* 9:147–164.

Hopfield, J.J., 1982: Neural networks and physical systems with emergent collective properties, *Proceedings of the National Academy of Sciences USA* 79:2554–2558.

Kirkpatrick, S., C.D. Gelatt, Jr., and M.P. Vecchi, 1983: Equations of state calculations by fast computing machines, *Journal of Chemical Physics* 21:1087–1091.

Maynard-Smith, J., 1970: Natural selection and the concept of a protein space, *Nature* 255:563–564.

Metropolis, N., A. Rosenbluth, M. Rosenbluth, A. Teller, and E. Teller, 1953: Equations of state calculations by fast computing machines, *Journal of Chemical Physics* 21:1087–1091.

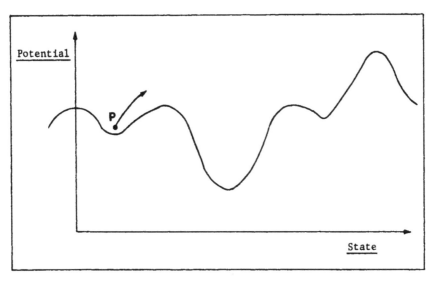

FIGURE 7.1 The current state of the system, represented here by the point P, has a finite probability of escaping from the local minimum at non-zero temperature. This combines the advantages of gradient- and random-search dynamics.

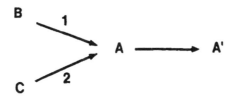

FIGURE 7.2 A four-neuron segment of the net

8. Genes as Bits for Nervous System Development

Introduction

No aspect of development places as great an informational load on the genome as does the development of the nervous system. The cells of the brain make up a system of awesome complexity, not due simply to the large number of cells involved, but to the high degree of individual identity in each cell relative to other tissues or organ systems. It is this high degree of individuality that makes possible the network circuitry of neuronal synapses, and the concomitant ability to process and transmit information that distinguishes neuronal function. In contrast to the immune system, the only other situation in which a comparable degree of complexity and diversity must be generated, the diversity among neurons depends as much on position and shape as on biochemical phenotype. Consequently, the genetic strategies are likely to differ.

The ontogeny and final form of the nervous system is species specific, with more in common the more closely related two species are, and is therefore largely under genetic control. Likewise, the variations in brain ontogeny between species reflects genetic variation. The nature of this genetic variation is one of the more intriguing questions of evolution, in that it led to the production of the enormously complex brains of the higher primates. It becomes one of the more intriguing questions of development from the standpoint that this increasing complexity arose without any increase in the size of our genome relative to our early mammalian ancestors.

Any attempt at constructing hypotheses concerning genetic strategies in nervous system development requires, as a minimal starting point, the defining of the informational units or genetic "bits" which take part in the process. That is, although we may know that some number of genes take part, this implies no particular strategy because of the enormous variety of possible gene functions. By analogy to the study of sensory physiology in the central nervous system, it was all very well to know that neurons were involved, but far more important to learn what categories of possible neuronal function there were as a prelude to building circuit models (e.g., Hubel and Wiesel 1962). Stated in an alternative way, the question of what are the genetic bits in nervous system ontogeny can be construed as defining how the genes divide up the developmental process. The question departs from conventional notions of informational units in that these genetic bits are in no sense uniform or homogeneous.

Developmental Genetic Systems

As an approach to this question, it is instructive to look at the examples we have at hand of such genes. Our information comes from studies of developmental mutations in the four eukaryotic organisms for which developmental mutants and the ability to analyze them are available: the nematode *Caenorhabditis elegans*, the fruit fly *Drosophila melanogaster*, the zebra fish *Brachydanio rerio*, and the house mouse *Mus musculus*. It is similarly useful to look at the genome size in relation to the neural complexity of each organism. (See Table 8.1.)

Several notable aspects emerge from this comparison. The leap in brain complexity does not correlate closely with the increase in non-neuronal cell complexity, with number of genes, and particularly not with total genome size. The clear trend is toward greater genetic efficiency, or perhaps versatility, in brain development.

These organisms also represent a progression in the relative plasticity in the development of their nervous systems. The nematode nervous system is by far the most "hard wired" in the sense of having an invariant number and pattern of cells, derived by means of strict cell lineages (Sulston et al. 1983), with little or no plasticity or experience dependence in the final cellular patterns (White et al. 1976). The fly is intermediate, and, in some respects, may be an actual mixture. The nerve cord of the larval nervous system may develop as rigidly as the nematode's (Thomas et al. 1984), whereas the cerebral hemispheres of the larva—and particularly the adult brain—are a great deal more complex, showing little or no strict cell lineage correlation with final patterns (Kankel and Hall 1976; Greenspan et al. 1980) and exhibiting some experience-dependent components (Technau 1984).

The fish and the mouse, to the extent that we know, are further extensions of this trend away from invariance and strict lineage-dependent patterns (Mullen 1977; Goldowitz 1986), and exhibit a more substantial degree of plasticity and experience dependence (Tsukahara

1981). (It is important to bear in mind that our knowledge of fly, fish, and mouse development is less comprehensive than of the nematode, and these conclusions may be overstatements due to insufficient data.)

Genetic Strategies and Evolutionary Changes

Given this context, what examples are there of mutations affecting nervous system development, and what clues do they provide into the questions of how the information for this aspect of ontogeny is stored and read out, and how these genetic strategies might have been able to vary in such a way as to produce evolutionary changes? Here, it is even more important to realize that we are dealing with a very small sample of the genes that are likely to be involved, simply because the rest have yet to be identified.

There appear to be four major classes of such genes, to a first approximation. The first group consists of genes that act in a hier-archical manner, mediating major binary choices of developmental pathways. This class is best exemplified by genes identified in nematodes and flies mediating sex determination (Cline 1984; Hodgkin and Bren-ner 1977), and by several loci called "homeotic" which channel the cells of whole regions of the organism into one pathway versus another (Lewis 1978; Wakimoto and Kaufman 1981; Greenwald et al. 1983). The effects of these genes on neural development are a result of the fact that when a whole region of the organism is being channeled into, for instance, the appropriate segmental identity, many of the neurons follow suit (e.g., Teugels and Ghysen 1985).

An interesting variant on this type of gene—one pertaining directly to the determination of neurons—is a set of loci in the fly which affect the decision of embryonic cells as to whether to become neural or epidermal (Lehmann et al. 1983). When mutant, most of the cells that would normally have become dermatoblasts become neuroblasts instead. In at least one of these instances, the switch clearly acts in a highly localized fashion (Hoppe and Greenspan 1986). Although there are no firm examples of this class in the mouse, there is good reason to think there will be analogs to the neural-epidermal switch genes, since the developmental relationship between these cell types is intimate in phyla as distant as coelenterates and mammals. In fact, there is a candidate gene in the mouse, one in which mutations transform non-neural ec-todermal tissue into ectopic neural tubes in the affected embryos (Gluecksohn-Schoenheimer 1949; Jacobs-Cohen et al. 1984; Greenspan and O'Brien 1986).

A second class of genes are those which can be roughly described as governing pattern globally in the organism, based exclusively on fly mutants that alter segmentation patterns or dorso-ventral patterns (Nusslein-Volhard 1979; Nusslein-Volhard and Wieschaus 1980). These affect neurons in the same way as the first class, in that the neurons follow suit when their segment is changed (Jan and Jan 1982; Hoppe and Greenspan, unpublished observations).

These first two classes have in common the characteristics that they are pleiotropic (affect many tissue and cell types), and they are probably not free to vary much without dire consequences. The pleiotropy in this case will be direct, the action of the gene directly on many cell types. The proposed inflexibility with respect to variation is likely to be due to the need for insulation of these steps in development from significant fluctuation, lest the organism lack essential structures.

A third class common to all four organisms is that governing cell-type-specific differentiation. These mutations eliminate particular neuronal cell types, with no apparent direct effects on others. Examples in the nematode include mutations affecting mechanosensory cells (Chalfie and Sulston 1981), or those affecting neurons involved in egg laying (Trent et al. 1983). Fly counterparts include the "small optic lobes" mutation, which eliminates a specific subset of columnar neurons in the optic lobes (Fischbach and Heisenberg 1981), and the "retinal degeneration" mutation, which eliminates a class of photoreceptor cells from each ommatidia of the compound eye (Harris and Stark 1977). On a larger scale, the "ventral-nervous-system-degeneration" mutant (White 1980) eliminates the entire ventral nerve cord in the fly embryo.

The zebrafish mutant *ndg-1* (neural degeneration; Grunwald et al. 1987) exerts its effects on CNS neurons as opposed to peripheral ones, but only on certain classes in the CNS. The zebrafish mutant B-39 (Grunwald and Streisinger 1987) specifically causes muscle disorganization. Many of the neurological mutants of the mouse (Sidman et al. 1965) fall into this class, such as "weaver," which lacks cerebellar granule cells (Rakic 1975), and "purkinje-cell-degeneration" (Mullen et al. 1976).

Most striking is the relative ease of isolating these mutations. One might explain their abundance on the grounds of greater viability as compared to the other kinds of mutations, a factor certainly relevant to the neurological mutants of the mouse, all of which must have been viable in order for the phenotype to be discernible and the strain propagated. This caveat does not apply, however, to the other organisms, in which lethal mutations are much easier to handle.

Another plausible explanation comes from correlating these genetic findings with biochemical studies of cell differentiation, in which a large (and seemingly endless) number of cell-type-specific growth and differentiation factors have been identified (Berg 1984). The best known of these, nerve growth factor, is required for sympathetic and sensory neuron survival (Berg 1984). An analogous factor involved in motor neuron survival appears to exist as well (Calof and Reichardt 1984). Similar factors have been identified for CNS neurons (Berg 1984), as well as for glial cells (Brockes et al. 1980). Similarly, the neuronal-adhesion molecules belong with these other factors (e.g., Edelman 1984; Rathjen and Schachner 1984). It is likely that many class II mutations will turn out to be defects in the production or response to such factors.

This class many be somewhat more prone to variation than the first two, in the sense that the components are limited to one or a few cell types in their sphere of action and, therefore, are somewhat less constrained. However, the freedom of variation will still be constrained for gene products such as growth factors by the need for correct molecular matching between receptors and effectors. Furthermore, the diversity of biochemical phenotypes available to neurons will be a function of this class, and although mammals certainly have a greater number than nematodes, the differences between nematodes and flies are likely to be minimal, and those between mice and humans negligible. Thus, a significant amount of the variation must lie elsewhere.

Pleiotropy and Evolutionary Plasticity

The fourth class of genes is the hardest to document with proven examples, but is one whose existence is strongly predicted by what we know from experimental embryology to be the high degree of interaction between cells and tissues, and from considerations of where evolutionary flexibility in the genome is likely to arise. These "interactive" genes should have many of the opposite properties from the previous categories. They should not be highly insulated from perturbations, they should exert their effects on many cell types in a pleiotropic fashion, and the consequences of variations in them should be more permissible.

The lack of insulation would be due to their taking part in many of the developmental interactions known to occur. Their pleiotropy would be of a more indirect variety than that of the first category. Here, changes would ripple through the system rather than impinging directly on many places, playing upon the nonlinearity of the process. The permissibility of their variation could be due to exploitation of the plasticity of the developmental system, and they, in turn, would further contribute to that plasticity (cf. Katz 1983). One prediction of this hypothetical class is its virtual absence in nematodes, its limited presence in flies, and its prominence in fish and particularly in mice.

Indeed, there are examples of pleiotropic developmental mutants in all of these organisms (e.g., Trent et al. 1983; Wright 1970; Gluecksohn-Waelsch 1951; Grunwald and Streisinger 1987), but whether their pleiotropy is direct or indirect has not been determined for the most part.

It is easy to imagine that this last class would contain genes that participate in multiple roles. For example, an enzyme involved in a basic metabolic pathway might also take part in a particular developmental step if its product were required for differentiation or as a signalling molecule at a certain time and place (McMahon 1974; Tomkins 1975). Cyclic AMP is an example of such a case, although an extreme one to the extent that it seems to participate in everything from pattern formation (Brenner 1977) to channel modulation (Kennedy 1983). One possible candidate for this class is the "brachymorphic" mutation in the mouse (Orkin et al. 1976), where a defect in sulfate metabolism produces a variety of pleiotropic effects, one of which is foreshortening of bones. The etiology of the defect stems from the requirement for sulfation of proteoglycans in the growing bone.

In addition to metabolites, there are also examples of structural proteins that perform multiple functions. The basal lamina protein laminin, for instance, has been shown in vitro to mediate hepatocyte spreading (Timpl et al. 1983), epidermal cell attachment (Terranova et al. 1980), and neurite extension (Edgar et al. 1984; Lander et al. 1985). It is a very large protein, has binding sites for a variety of extracellular proteins and molecules, and may represent a kind of multifunctional gene capable of picking up or losing roles as the consequence of minor changes in structure.

The wide spectrum of cell surface proteins revealed by immunological studies—some of which exhibit cellular distributions that defy comprehension—may also contain potential members of this class of genes. The most intensely studied examples, products of the H-2 locus—the major histocompatibility complex of the mouse—are present on virtually all cell types, yet the only functions defined so far are those pertaining to lymphocytes and macrophages in the immune response (Klein et al. 1981). Neuronal surface proteins have also been identified in this way, many of which are present on other cell types as well, but not in any consistent pattern (e.g., Boyse and Old 1976; Matthew et al. 1985). An interesting interpretation of the existence of such distinct yet overlapping sets is that they are components of a combinatorial system specifying the differentative pathway a cell follows by dint of the combination of these genes that it expresses (cf. Boyse and Old 1976). One of these proteins, coded by the *Thy-1* locus of the mouse, has been implicated in neuronal development from in vitro studies of neurite extension (Leifer et al. 1984; Greenspan and O'Brien 1987).

Also in this class would be changes in parameters such as the number of cells proliferated to form the founder cell pool for a given structure. The critical factors could be rather subtle and not necessarily direct, requiring perhaps very mild, quantitative genetic changes in activities such as the expression of cellular oncogenes (e.g., Campisi et al. 1984; Fredericksen and McKay 1987). Moreover, they could have great consequences for the final structures and patterns formed, especially in the brain, where this sort of variation may have been instrumental in the enormous expansion of brain size and structure accompanying vertebrate evolution.

An interesting example of this kind of variation is provided by the evolution of multiple sensory field maps during the course of mammalian phylogeny. Lower mammals such as hedgehogs have been found to have two complete topographical maps of visual and somatosensory modalities in their neocortex, while monkeys have 10 or more discrete representations of each (Merzenich and Kaas 1980). Such changes are not likely to involve massive genetic reprogramming, nor any change in the number of biochemical phenotypes available to neurons. Rather, they may simply involve subtle alterations which, in the context of the existing gene sets, give multiple versions of a group of cells where only one existed before. That is, mechanisms for the formation of coherent nerve networks may operate on whatever founder pool they are given in such a way that when the pool exceeds a certain size, they do not continue simply to make larger versions of the same structure, but instead duplicate them. This kind of scenario could plausibly account for much of the explosive forebrain development found in vertebrate evolution.

Disruptive as it may sound, there are examples of developmental mechanisms capable of buffering what would otherwise be chaotic, and presumably lethal, changes in neuron number (cf. Katz 1983). Some of these mechanisms have been revealed in the course of embryological experiments involving transplantation and subsequent orderly growth and differentiation of excessive amounts of neural tissue (Constantine-Paton 1982). In such cases, the final pattern is far from chaotic, and appears to be governed by a mixture of relative preferences in adhesiveness between cells (cf. Edelman 1984; Bonhoeffer and Huf 1985) and a sorting-out process dependent on the electrical activity (i.e., functional use) of neighboring cells. There is already ample evidence that functional activity plays a role in neuron survival, as well as in the modification of neuronal connections in development (cf. Harris 1981).

The designation of four classes of genes for neural development will be subject to revision as more genes are identified and their characteristics detailed. It may turn out, for instance, that there is little essential difference between the first two classes, that they are merely varieties of the set of genes charged with setting up and determining the major subdivisions of the organism's global pattern. It is entirely possible, however, that vertebrate homologs to such genes do not exist, DNA homologies notwithstanding (McGinnis et al. 1984), and that different genetic strategies are called into play. It is equally possible that some class I genes will turn out to be switches mediated by multifunctional class IV-type genes. This might obtain, for example, if a determinative decision was triggered by a cell's contact with appropriate components of the extracellular matrix.

It seems unlikely (although not impossible, cf. Tomlinson and Ready 1985) that the third class will spill over into the others, comprised as it is of cell-type-specific genes. The fourth class is also unlikely to overlap with the first three, being neither cell-type-specific nor strictly hierarchical. Multifunctional, pleiotropic genes cut across all simple boundaries, by definition, and thus are the most likely vehicles for embodying the nonlinear interactions in development.

Implicit in the designation of these classes is that much of what governs neural development does not act exclusively on the nervous system. Only class III is highly specific, and although it may turn out that there are a very large number of loci in this category, due to the spectrum of neuronal phenotypes, they are not likely to be the major source of evolutionary flexibility that nervous systems show. With respect to the first two classes, there are no examples, nor any hints, of analogous patterning loci which act exclusively on the nervous system. This is in keeping with the observation that none of the genes in this group respect boundaries of cell type.

Genes of the fourth class might be expected to have effects in the nervous system they do not have elsewhere, but not necessarily due to their lack of expression elsewhere. It is these genes that emerge from this discussion as the most crucial in accounting for the origin of complexity in the nervous system. They are the most ideally suited for producing and responding to subtle, quantitative variations. Such variations provide the most plausible source for the wide ranges of brain development, while at the same time, they are the raw material of population genetics and natural selection.

Bibliography

Berg, D., 1984: New neuronal growth factors, *Annual Review of Neuroscience* 7:149–170.

Bonhoeffer, F., and J. Huf, 1985: Position-dependent properties of retinal axons and their growth cones, *Nature* 315:409–410.

Bonner, J.T., 1987: *The Evolution of Complexity*, Princeton University Press, Princeton, N.J. (in press).

Boyse, E.A., and L.J. Old, 1976: The immunogenetics of differentiation in the mouse, Harvey Lectures series 71:23–54.

Brenner, M., 1977: Cyclic-AMP gradient in migration pseudoplasmodium of the cellular slime mold *Dictyostelium discoideum, Journal of Biological Chemistry* 252:4073–4077.

Brockes, J.P., G.E. Lemke, and D.R. Balzer, 1980: Purification and preliminary characterization of a glial growth factor from the bovine pituitary, *Journal of Biological Chemistry* 255:8374–8377.

Calof, A.L., and L.F. Reichardt, 1984: Motorneurons purified by cell sorting respond to two distinct activities in myotube conditioned medium, *Developmental Biology* 106:194–210.

Campisi, J., H.E. Gray, A.B. Pardee, M. Dean, and G.E. Sonenenshein, 1984: Cell-cycle control of c-*myc* but not c-*ras* expression is lost following chemical transformation, *Cell* 36:241–247.

Chalfie, M., and J. Sulston, 1981: Developmental genetics of the mechanosensory neurons of *Caenorhabditis elegans, Developmental Biology* 82:358–370.

Cline, T.W., 1984: Autoregulatory functioning of a *Drosophila* gene product that establishes and maintains the sexually determined state, *Genetics* 107:231–277.

Constantine-Paton, M., 1982: The retinotectal hookup—the process of neural mapping, in S. Subtelny and P.B. Green (eds.), *Developmental Order: Its Origin and Regulation*, Alan R. Liss, Inc., New York, 317–349.

Edelman, G.M., 1984: Modulation of cell adhesion during induction, histogenesis, and perinatal development of the nervous system, *Annual Review of Neuroscience* 7:339–377.

Edgar, D., R. Timpl, and H. Thoenen, 1984: The heparin-binding domain of laminin is responsible for its effects on neurite outgrowth and neuron survival, *EMBO Journal* 3:1463–1468.

Fischbach, H.K., and M. Heisenberg, 1981: Structural brain mutant of *Drosophila melanogaster* with reduced cell number in the medulla cortex and normal optomotor response, *Proceedings of the National Academy of Sciences USA* 78:1105–1109.

Fredericksen, K., and R.D.G. McKay, 1987: Proliferation and differentiation of neuroepithelial precursor cells, *Journal of Neuroscience* (in press).

Gluecksohn-Waelsch, S., 1951: Physiological genetics of the mouse, *Advanced Genetics* 4:1–51.

Gluecksohn-Schoenheimer, S., 1949: The effects of a lethal mutation responsible for duplications and twinning in mouse embryos, *Journal of Experimental Zoology* 110:47–76.

Goldowitz, D., 1987: A "clonal" analysis of somatosensory cortex barrel formation using chimeric mice (submitted for publication).

Greenspan, R.J., J.A. Finn, and J.C. Hall, 1980: Acetylcholinesterase mutations in *Drosphila* and their effects on the structure and function of the central nervous system, *Journal of Comparative Neurology* 189:741–774.

Greenspan, R.J., and M.C. O'Brien, 1986: Genetic analysis of mutations at the fused locus of the mouse, *Proceedings of the National Academy of Sciences USA* 83:4413–4417.

Greenspan, R.J., and M.C. O'Brien, 1987: Genetic evidence for *Thy-1*'s role in neurite outgrowth in the mouse (submitted for publication).

Greenwald, I., P.W. Sternberg, and H.R. Horvitz, 1983: The *lin-12* locus specifies cell fates in *Caenorhabditis elegans, Cell* 34:435–444.

Grunwald, D.J., and G.D. Streisinger, 1987: Frequency of specific locus and recessive lethal mutations induced by ethyl nitrosourea in zebrafish, *Brachydanio rerio* (submitted for publication).

Grunwald, D.J., M. Westerfield, G. Streisinger, and C.B. Kimmel, 1987: A neural degeneration mutation that spares primary neurons in the zebrafish (submitted for publication).

Harris, W.A., 1981: Physiological activity and development, *Annual Review of Physiology* 43:689–710.

Harris, W.A., and W.S. Stark, 1977: Hereditary retinal degeneration in *Drosophila melanogaster*—a mutant defect associated with the phototransduction process, *Journal of General Physiology* 69:261–292.

Herman, R.K., and H.R. Horvitz, 1980: Genetic analysis of *Caenorhabditis elegans*, in B.M. Zuckerman (ed.), *Nematodes as Biological Models*, Academic Press, New York, 228–261.

Herskowitz, I.H., 1950: An estimate of the number of loci in the X chromosome of *D. melanogaster, American Naturalist* 84:255–260.

Hodgkin, J.A., and S. Brenner, 1977: Mutations causing transformation of sexual phenotype in the nematode *Caenorhabditis elegans, Genetics* 86:275–287.

Hoppe, P.W., and R.J. Greenspan, 1986: Local function of the Notch gene in embryonic ectodermal pathway choice during *Drosophila* embryogenesis, *Cell* 46:773–783.

Hubel, D.H., and T.N. Wiesel, 1962: Receptive fields, binocular interaction, and functional architecture in the cat's visual cortex, *Journal of Physiology* 160:106–154.

Jacobs-Cohen, R.J., M. Speigelman, J.C. Cookingham, and D. Bennett, 1984: Knobbly, a dominant mutation in the mouse that affects embryonic ectoderm organization, *Genetical Research (Cambridge)* 43:43–50.

Jan, L.Y., and Y.N. Jan, 1982: Antibodies to horseradish peroxidase as specific neuronal markers in *Drosphila* and grasshopper embryos, *Proceedings of the National Academy of Sciences USA* 79:2700–2704.

Kankel, D.R., and J.C. Hall, 1976: Fate mapping of nervous system and other internal tissues in genetic mosaics of *Drosophila melanogaster, Developmental Biology* 48:1–24.

Katz, M.J., 1983: Ontophyletics—studying evolution beyond the genome, *Perspectives on Biological Medicine* 26:323–333.

Kennedy, M.B., 1983: Experimental approaches to understanding the role of protein phosphorylation in the regulation of neuronal function, *Annual Review of Neuroscience* 6:493–525.

Klein, J., J. Juretic, C.N. Baxevanis, and Z.A. Nagy, 1981: The traditional and a new version of the mouse H-2 complex, *Nature* 291:455–460.

Lander, A.D., D.K. Fujii, and L.F. Reichardt, 1985: Laminin is associated with the "neurite outgrowth-promoting factors" found in cultured media, *Proceedings of the National Academy of Sciences USA* 82:2183–2187.

Leifer, D., S.A. Lipton, C.J. Barnstable, and R.H. Masling, 1984: Monoclonal antibody to Thy-1 enhances regeneration of processes by rat retinal ganglion cells in culture, *Science* 224:303–306.

Lehmann, R., F. Jimenez, U. Dietrich, and J.A. Campos-Ortega, 1983: On the phenotype and development of mutants of early neurogenesis in *Drosophila melanogaster, Wilhelm Roux' Archives of Developmental Biology* 192:62–74.

Lewis, E.B., 1978: A gene complex controlling segmentation in *Drosophila, Nature* 276:565–570.

Matthew, W.D., R.J. Greenspan, A.D. Lander, and L.F. Reichardt, 1985: Immunopurification and characterization of a neuronal heparan sulfate proteoglycan, *Journal of Neuroscience* 5:1842–1850.

McGinnis, E., C.P. Hart, W.J. Gehring, and F.H. Ruddle, 1984: Molecular cloning and chromosome mapping of a mouse DNA sequence homologous to homeotic genes of *Drosophila, Cell* 38:675–680.

McMahon, D., 1974: Chemical messengers in development—a hypothesis, *Science* 185:1012–1021.

Merzenich, M.M., and J.H. Kaas, 1980: Principles of organization of sensory-perceptual systems in mammals, *Progress in Psychobiology and Physiological Psychology* 9:1–42.

Mullen, R.J., 1977: Site of *pcd* gene action and Purkinje cell mosaicism in cerebella of chimaeric mice, *Nature* 270:245–247.

Mullen, R.J., E.M. Eicher, and R.L. Sidman, 1976: Purkinje cell degeneration, a new neurological mutation in the mouse, *Proceedings of the National Academy of Sciences USA* 73:208–212.

Muller, H.J., and E. Altenberg, 1919: The rate of change of hereditary factors in *Drosophila, Proceedings of the Society for Experimental Biology* 17:10–14.

Nusslein-Volhard, C., 1979: Maternal effect mutations that alter spatial coordinates of the embryos of *Drosophila melanogaster*, in S. Subtelny and I.R. Konigsberg (eds.), *Determinants of Spatial Organization*, Academic Press, New York, 185–211.

Nusslein-Volhard, C., and E. Wieschaus, 1980: Mutations affecting segment number and polarity in *Drosophila, Nature* 287:795–801.

Orkin, R.W., R.M. Pratt, and G.R. Martin, 1976: Undersulfated chondroitin sulfate in the cartilage matrix of brachymorphic mice, *Developmental Biology* 50:82–94.

Rakic, P., 1975: Synaptic specificity in the cerebellar cortex—study of anomalous circuits induced by single gene mutations in mice, *Cold Spring Harbor Symposium on Quantitative Biology* 40:333–346.

Rathjen, F.G., and M. Schachner, 1984: Immunocytological and biochemical characterization of a new neuronal cell surface component (L1 antigen) which is involved in cell adhesion, *EMBO Journal* 3:1–10.

Sidman, R.L., M.C. Green, and S.H. Appel, 1965: *Catalog of the Neurological Mutants of the Mouse*, Harvard University Press, Cambridge, Mass.

Strausfeld, N.J., 1976: *Atlas of an Insect Brain*, Springer-Verlag, Berlin.

Sulston, J.E., E. Schierenberg, J. White, and N. Thomson, 1983: The embryonic cell lineage of the nematode *Caenorhabditis elegans*, *Developmental Biology* 100:64–119.

Technau, G.M., 1984: Fiber number in the mushroom bodies of adult *Drosophila melanogaster* depends on age, sex, and experience, *Journal of Neurogenetics* 1:113–126.

Terranova, V.P., D.H. Rohrbach, and G.R. Martin, 1980: Role of laminin in the attachment of PAM 211 (epithelial) cells to basement membrane collagen, *Cell* 22:719–726.

Teugels, E., and A. Ghysen, 1985: Domains of action of *bithorax* genes in *Drosophila* central nervous system, *Nature* 314:558–561.

Thomas, J.B., M.J. Bastiani, M. Bate, and C. Goodman, 1984: From grasshopper to *Drosophila*—a common plan for neuronal development, *Nature* 304:440–442.

Timpl, R., S. Johansson, V. van Delden, I. Oberbaumer, and M. Hook, 1983: Characterization of protease-resistant fragments of laminin mediating attachment and spreading of rat hepatocytes, *Journal of Biological Chemistry* 258:8922–8927.

Tomkins, G.M., 1975: The metabolic code, *Science* 189:760–763.

Tomlinson, A., and D.F. Ready, 1986: *Sevenless*—a cell-specific homeotic mutation of the *Drosophila* eye, *Science* 231:400–402.

Trent, C., N. Tsung, and H.R. Horvitz, 1983: Egg-laying defective mutants of the nematode *Caenorhabditis elegans*, *Genetics* 104:619–647.

Tsukahara, N., 1981: Synaptic plasticity in the mammalian central nervous system, *Annual Review of Neuroscience* 4:351–379.

Wakimoto, B.T., and T.C. Kaufman, 1981: Analysis of larval segmentation in lethal genotypes associated with the Antennapedia gene complex in *Drosophila melanogaster*, *Developmental Biology* 81:51–64.

White, J.G., E. Southgate, J.N. Thomson, and S. Brenner, 1976: The structure of the ventral cord of *Caenorhabditis elegans*, *Philosophical Transactions of the Royal Society of London B* 275:327–348.

White, K., 1980: Defective neural development in *Drosophila melanogaster* embryos deficient for the tip of the X chromosome, *Developmental Biology* 80:332–344.

Wright, T.R.F., 1970: The genetics of embryogenesis in *Drosophila*, *Advanced Genetics* 15:261–395.

	Nematode
Haploid DNA content (base pairs)	1.5×10^8
Lethally mutable genes*	2,000
Total genes[†]	2,500
Non-neuronal cell types*	20
Neurons in brain	300

(*These values are estimates, [†] these ar

Sources: Bonner, 1987; Sulston, et al
Herman & Horvitz, 1980; Muller & Alte
Herskowitz, 1950; Strausfeld, 1976; G
Streisinger, 1987; D.J. Grunwald, per
W.D. Dove, personal communication.)

9. Dynamics of Image Formation by Nerve Cell Assemblies

The Bases of Neural Cooperativity

Certain general attributes of the brain enable it to generate large-scale patterns of cooperative activity. These include the immense numbers of nerve cells packed at high density; the filamentous threads of protoplasm that each neuron extends over distances relatively great with respect to the diameters of neurons; the lack of protoplasmic contact between neurons that gives each a significant degree of autonomy; and the specialized junctions or synapses that neurons sustain with thousands of others within their arbors, such that there are innumerable feedback loops. Each neuron has a relatively narrow near-linear range of function and a relatively broad nonlinear dynamic range. In the appropriate circumstances of excitation, regenerative domains of action arise that organize vast numbers of neurons transiently into macroscopic space-time patterns of collective activity. These ultimately determine the patterned firings of motor neurons that give rise to what we experience and observe as goal-directed behavior. My intent here is to describe some basic biophysical properties of those cooperative actions in the brain.

By analogy, these events may resemble some well-known macroscopic phenomena observed in complex, diffusion-coupled chemical reactions. These systems share with the brain the attributes of large numbers of semi-autonomous microscopic particles, widespread weak interactions over closed-loops paths, delays in transmission, nonlinear input-output

Reprinted with permission from _Synergetics of the Brain_, Proceedings of the International Symposium on Synergetics at Schloss Elmau, Bavaria, 2–7 May 1983. Springer-Verlag © 1983, pp. 102–121.

relations for the particles, predominance of elastic over reactive collisions, and a separation of time and distance scales for the microscopic and macroscopic events.

There are important differences (Freeman 1975, 1981). The active states of neural systems refer to informational transactions that are measured ultimately with respect to the strengths of sensory stimuli and of muscle contractions and not to energy standards. The neural activity is carried in electrical currents, chemical fluxes, and the like, but the first and second laws of thermodynamics do not constrain higher brain function. In respect to its metabolic needs for information processing, the brain has an unlimited supply of energy and an unlimited sink for waste heat and entropy. Neural activity is not conserved, nor is it subject to relentless decay.

Moreover, the brain is not limited to diffusion processes. These do occur in the form of so-called electrotonic current fields, in synaptic clefts, and in the extracellular medium of the brain, but the nerve filaments and their specialized junctions provide relatively high-speed, point-to-point transmission over great distances. This implies that transmission delays are not necessarily distance-dependent, as with diffusion, and that spatially extended transmission need not undergo the smoothing action of diffusion.

Concerning structure, on the one hand, there is a stability and complexity for the substrate of cooperative behavior in the nervous system that is not found in the test tube or Petri dish. One can treat a chemical system as homogeneous to begin with, but one cannot consider brain function seriously without a detailed knowledge of neuroanatomy. On the other hand, whereas the boundary conditions are crucial for the solution of equations describing chemical systems, for the nervous system, they are essential but trivial. In a functional sense, the brain is not spatially bounded, and active regions can be represented by closed geometries such as a toroid for an active focus in a sheet of nerve cells.

State Variables of Microscopic and Macroscopic Activity

Neural activity at the microscopic level of single neurons occurs in two main forms, the pulse and the wave, that are characteristically associated with two main types of filament: the dendrite and the axon (Figure 9.1). At the macroscopic level of the cooperative ensemble of neurons, the state variables take different forms. The ensemble state variables are the pulse density, which is the sum of pulses at any moment over a local neighborhood in an ensemble, and the wave density, which is the instantaneous sum of dendritic currents over the neurons forming that local neighborhood.

The electrical currents that sustain neural pulse and wave activity have complex geometries. In most circumstances, neither the membrane sites of electromotive force nor the sites where the currents act on membrane are directly accessible. Potential differences are measured at convenient locations in or outside of neurons. For these and other reasons, the observable potentials are not identical to the actions of currents. Much of the work of neurophysiology consists of constructing transforms to derive estimates of active states from their observable electrical manifestations. These aspects are dealt with elsewhere (Freeman 1975, 1980).

The state variables of single neurons are accessed with relative ease by use of the microelectrode for recording transmembrane potential in the wave mode and the unit potential in the pulse mode. Access to statistical averages of wave and pulse activity is more difficult. The first step is to ensure that an assembly has been properly defined. Proper definitions are based on connectivity and dynamic relations, not on temporal or spatial contiguity; distinct ensembles are usually commingled within cortex (Freeman 1975). In some circumstances that allow one to assume stationarity, one can invoke the ergodic hypothesis and construct averages over a collection of recordings from single neurons within a neural assembly. For nonstationary events, it is essential to record the activities of large numbers of neurons simultaneously and measure their sums.

It is technically not feasible at present to derive a sufficiently large number of unit pulse trains to represent the state variables of an ensemble in the pulse mode experimentally. It is feasible to do this for the wave mode by virtue of the fact that dendritic currents occur in closed loops, with an electromotive force in the dendritic membrane, a site of action of the current in the axonal membrane, and a return current path in the extracellular tissue space. The electrical potentials established by the currents over the tissue resistance add scalarly to give the sum of dendritic currents over local neighborhoods in an ensemble. These are commonly called electroencephalographic (EEG) waves. In the proper experimental conditions, an estimate of the state variable of one ensemble can be derived and isolated among the overlap of contributions from many ensembles (the "cocktail party effect").

There are two reciprocal operations between the two modes. At the microscopic level, the dendrites of each neuron receive pulses at synapses and convert them to continuously variable dendritic current. The flow of current across the axonal membrane modulates the axonal firing rate. For each neuron, there is a time-varying nonlinear relationship between pulse rate at the synapse and the amplitude of dendritic current (Figure 9.2). There is a time-varying linear relationship between axonal transmembrane potential established by dendritic current and axonal pulse firing rate between the limits of threshold and cathodal block. The main function of the dendrites is the weighted space-time integration of the local responses established by pulses, with convergence of the sum to the axon. The main function of the axon is translation and spatial divergence of the output, with multiplication of signal amplitude by the numbers of its branches. Those operations also invoke conduction and synaptic time delays. Further analysis of function at the level of single neurons proceeds by description of the connections of networks and by computer simulation with nonlinear network models.

In macroscopic ensembles, wave-pulse conversion by axons appears as a static sigmoidal nonlinearity that is asymptotic to zero pulse rate and to a maximum that is a property of the ensemble related to its mean firing rate. The derivative of this curve defines the axonal gain; changes in the derivative relating to arousal can be expressed in a multiplicative coefficient that also represents the mean level of background activity in the ensemble. Pulse-wave conversion is constrained within a static linear range by the wave-pulse nonlinearity owing to their sequential occurrence in loops and cascades. Because the ensemble pulse-to-wave conversion is linear, it can be represented by a coefficient that is fixed for any desired time period, but can be changed to represent synaptic changes with learning. This synaptic linearity greatly simplifies further analysis. The axonal and synaptic time delays within cooperative domains can be incorporated with comparable dendritic cable delays in one stage of temporal integration. Axonal and dendritic divergence properties similarly can be combined into a single stage of spatial integration.

The derivation of activity in the wave mode is most easily done for ensembles that are organized in layers, such as arrays of sensory receptors and the neurons that comprise the cortex of the cerebrum and cerebellum, as well as various laminated nuclei. Activity can then be expressed in 2-dimensional pulse and wave density functions to represent the state variables of ensembles. Further, there is anatomical and physiological evidence for the spatial coarse-graining of neural activity by what are known as cortical columns and glomeruli. Therefore, the representation of spatially distributed cortical events can be discretized into manageably small numbers of state variables, and the neural dynamics can be represented with coupled ordinary differential equations, which incorporate static nonlinear functions for wave-pulse conversion and coupling coefficients for pulse-wave conversion that depend on conditions imposed from outside the ensemble by other parts of the brain.

Experimental analysis has shown that the parameters representing delay and divergence can be treated as invariant over both time and space and for the olfactory system over ensembles. These facts also greatly facilitate modeling (Freeman 1975) by restricting the assignment of behavioral-state-dependent changes in the model to synaptic gains (for attention and habituation) and to axonal gains (for arousal).

The function of a neural layer is analyzed by estimation of the space-time patterns of the distributed input and output state variables. A set of equations is constructed to represent the neural dynamics, with input and initial conditions that simulate the neural stimulus and background noise. The solutions simulate the observed neural output. The equations then serve formally to define the operations of the neural layer. Because there is no energy standard for neural active states, it is essential to begin with sensory input. Most sensory stimuli are blocked early by neural mechanisms of attention and habituation, so it is preferable to begin with a stimulus that elicits a behavioral response. The existence of neural information processing is thereby assured. Much of the neural activity of the brain is background noise that serves to maintain its various parts in optimal and often near-linear ranges. The establishment of initial conditions requires characterization and measurement of this noise. The test of a model with real data requires that time functions be calculated and compared with those estimated experimentally for the sensory input, and the neural activity output at each of an array of anatomical points over the neural surface, at spacings corresponding to the coarse graining of neural events.

Operations of Image Formation and Preprocessing

An example of these principles is afforded by a study of information processing in the mammalian olfactory system (Figure 9.3). An array of roughly 10^8 receptors is found in a sheet on the inner lining of the nose. Each receptor performs a transduction of a chemical concentration for particular odor or set of odors to a wave potential (manifested singly in a receptor potential and en masse in the electro-olfactogram) and then to a pulse frequency on its axon. In accordance with Johannes Muller's law of specific nerve energies, the rate of the train of pulses conveys information on intensity but not quality. Different receptors at different locations have different odor sensitivities. We infer that the receptors express the presence of each discriminable odor in a spatial pattern of pulse density over the array. The flow of olfactory input is time-parsed at the rate of respiration. There are no known mechanisms for interaction among receptors, so this ensemble is not interactive. I have called its representation a KO set (Freeman 1975).

The output is transmitted to the olfactory bulb by the olfactory nerve. This nerve has a degree of topographic order, so there is a conformal mapping of the receptor activity-density function onto the bulb (Figure 9.4). In addition, there is significant smoothing by temporal dispersion, owing to distributions in axonal conduction distances and velocities. In mammals, the receptors are activated (generally by excitation) only during inspiration of air through the nose. Then, the nerve input to the bulb is a smoothed surge of spatially patterned pulse density with each inspiration.

In brief, the first stage of olfaction in the nose and primary nerve performs the operations of chemotransduction, formation of a spatial image, gating or parsing of the flow of neural activity, and temporal smoothing within each time frame. With respect to time, it is a sampled-data system.

The receptor axonal endings are excitatory to the dendrites of an array of excitatory neurons in the bulb, the mitral cells, whose axons form the output path of the bulb, the olfactory tract. There are two main sets of interneurons in the bulb that perform diverse operations. The outer set is interactive by mutual excitation, and is represented by a KI_e set. Analysis has revealed the following properties. The set has two stable states, one of zero activity and the other of steady-state activity at a level dependent on the state of the bulb determined by centrifugal input from the brain. Excitatory perturbation from the nerve causes transient increase in interneuronal activity, with monotonic decay to the non-zero rest state. The decay rate is directly related to response amplitude, showing that the mechanism of stabilization is by saturation (Freeman 1975).

Output is in two forms. There is synaptic output to mitral cells that serves to maintain steady-state background activity in that population, and thereby, a linear dynamic range at rest. There is non-synaptic output that attenuates the strength of excitatory action of receptor axons onto bulbar neurons. This provides for dynamic input range compression, logarithmic conversion, and normalization of receptor input. The architecture, connections, and dynamics of these interneurons also provide for spatial and temporal integration and differentiation, including spatial contrast enhancement, local smoothing, coarse graining, holding, and gating, as well as bias control of the inner mechanism of the bulb. Hence, the image pre-processing operations that occur in sequential stages in other sensory systems are compressed into one synaptic stage in an outer layer of the bulb (Freeman 1975).

Formation of a Carrier Wave for Central Images

The mitral cells deliver excitation to each other. They form a mutually excitatory ensemble that is represented also by a KI_e set. This delivers excitation to the inner ensemble of interneurons, which is mutually inhibitory and is represented by a KI_i set. It inhibits the mitral ensemble, thereby forming a complex interactive ensemble that I have represented by a KII set (Figure 9.5). There are four types of forward synaptic gain in the KII set: excitation of excitatory neurons (k_{ee}) and of inhibitory neurons (k_{ei}), and inhibition of excitatory neurons (k_{ie}) and of inhibitory neurons (k_{ii}). There are three types of feedback gain: negative feedback ($k_{ei}k_{ie}$), mutual excitation (k^2_{ee}), and mutual inhibition (k^2_{ii}). There is no self-excitation or self-inhibition.

Analysis has shown the following properties. The open loop time constants of the population impulse responses on electrical stimulation conform to values predicated from the cable properties of the mitral and interneuron dendrites. The linear part of the dendritic operation can be approximated by an ordinary time-invariant, second-order differential equation (Figure 9.6), in which the low rate constant represents the RC properties of the membrane and the high rate constant represents the lumped serial conduction, synaptic, and dendritic cable delays, each of which conform to a 1-dimensional diffusion operation that is invariant under convolution (Freeman 1975).

The space constants in forward gains subserving mutual excitation parallel to the bulbar surface are large. This reflects the importance of this pathway for establishing coordinated activity over the surface of the bulb. The space constants for gains subserving mutual inhibition have not been measured directly, but must be at least as large, because there is no experimental evidence for travelling waves of excitation in the bulb other than pseudowaves imposed by afferent conduction delay. On the contrary, the bulbar waves are standing waves. The space constants for gains subserving negative feedback are minimal. This conforms to the anatomical substrate (the dendrodendritic reciprocal synapse between mitral and interneurons), and serves to prevent the development of chaos in bulbar activity. Oscillation develops in bulbar activity at a characteristic frequency of 40–80 Hz due to the negative feedback relation, and although it is modulated over time, it is instantaneously the same at all points over the bulbar surface of an active domain (Freeman 1975, 1979b).

The KII set has multiple stable states, of which three are of paramount importance. The lowest is zero activity, which is imposed on the bulb by deep anesthesia and is useful for measuring open-loop time and space constants. The next is a quasi-equilibrium state that is maintained by steady input from the KI_e set of external interneurons. The third is a limit cycle state with near-sinusoidal activity at a characteristic frequency of 40 to 80 Hz, with minor contributions from second and third pseudo-harmonics. Other higher stable states tend to be pathological, i.e., epileptiform (Babloyantz and Kaczmarek 1979).

The commonality of frequency is important with respect to the output of the bulb. The mitral cells send their axons in the olfactory tract to the olfactory cortex. The pulses on the axons appear to occur at random when observed one neuron at a time, seeming to conform to the output of a Poisson process with a short dead time (the refractory period). This confirms one important assumption in deriving the ensemble properties. However, when the probability of firing for single neurons is computed conditional on the amplitude of the amplititude of the dendritic potential for the local neighborhood, it is found to oscillate sinusoidally at the same frequency but with a pronounced phase shift. To the extent that neurons in the local neighborhood are in phase, the local subset transmits a continuous pulse density wave to the cortex that is an image of the local dendritic wave (Freeman 1975).

The tract differs from the nerve in having little detectable topographic organization, except in that part directed to the anterior olfactory nucleus. The activity from each local neighborhood in the bulb diverges widely over the cortex; conversely, each local neighborhood performs a running temporal and spatial integration over a broad region of the bulb. Because of the common frequency, the summation can be described as vectorial, and the effective input to the cortex is sinusoidal at that frequency. Indeed, the cortical peak frequency always occurs at the bulbar frequency when there is coherent activity in the bulb but not otherwise, showing that the cortex is driven by the mitral cells (Bressler and Freeman 1980). To the extent that frequency or phase dispersion occurs in the bulb, the signal transmitted to the cortex is degraded, and the cortical response amplitude is diminished (Bressler 1982).

Spatial Modulation of the Carrier Wave

The bulbar and cortical fields of potential both show the strong tendency for a brief burst of oscillation to occur during each inspiration (Figure 9.7). The mechanism of burst formation is of particular interest for an understanding of olfactory function. One factor relates to the central states of the brain: the animal must be aroused and in a motivated state. Another factor relates to air flow in the nose: if the nostrils are obstructed, there is no burst, even with mouth breathing. Granted that there are always background odors, bursts occur regularly whether or not a test odor is superimposed on the background odors in the air flow. The volley of action potentials from receptors during inspiration is smoothed, and does not drive the bulb at the frequency within the burst. Rather, it causes a slow wave response on which the burst is superimposed.

Simulation of the bulbar mechanism with the KII model shows that the burst results from a transition from the equilibrium state to the limit cycle state and back again with each inspiration. The two key features of the mechanism are the mutually excitatory connections and the nonlinear relation between wave amplitude and pulse density. This relation has been evaluated experimentally from the time-lagged pulse probability conditional on EEG amplitude. A function that closely describes it has been derived in part from the Hodgkin-Huxley equations (Figure 9.8). The derivative dP/dW of the function is a major determinant of the forward gain of local neighborhoods of neurons (Figure 9.9). The maximal gain is displaced to the excitatory side of the resting or equilibrium value for wave amplitude. When these neurons receive excitatory input, there is a coupled increase in their forward gain. Because they are interconnected by excitatory positive feedback, there is a regenerative increase in their feedback gain. This increases also the negative feedback gain. If the increase is sufficient, the bulb is driven out of its domain of equilibrium and into its limit cycle domain. It remains there until the excitatory input is withdrawn (Freeman 1979a,b).

This mechanism has been verified by applying long-acting excitatory drugs such as carbachol to the bulb and cortex to induce limit cycle activity of indefinite duration at the predicted frequency. (Limit cycle activity has also been induced by application of inhibitory blocking agents such as picrotoxin, but the frequency is much lower, as predicted.) In the early onset of drug action, the bursts occur only during inspiration, again showing the importance of the surge of nerve input for destabilization of the bulb (Freeman 1975).

From this consideration, three predictions were made: first, that in waking and motivated animals, there would be found a spatially inhomogenous pattern of burst amplitude over the bulbar surface conforming to axonal input from receptors sensitive to background odors; second, that the spatial pattern would change to a new pattern when a test odor was given to the animal; and third, that distinctive burst patterns would occur with each discriminable odor. EEG recordings were made by surgically implanting onto the bulb of rabbits arrays of 64 electrodes at spacings determined from spatial frequency analysis of the EEG. Indeed, a distinctive spatial pattern was found for each subject, which varied slightly in form over successive bursts without test odors, but the differences with test odors did not exceed those found for the background (Freeman 1978).

Changes in burst spatial pattern did occur when animals were conditioned to respond to a test odor by pairing each presentation with a brief electrical shock or a reward. The difference between bursts with and without the test odor appeared during the first or second session with each new odor as the conditioned response emerged, but in subsequent sessions with that odor, the burst pattern stabilized in a new form and persisted whether or not the test odor was actually present. The conclusion was drawn that the burst pattern was determined more by internal factors in the bulb than by input patterns (Freeman and Schneider 1982).

These results were consistent with findings by several investigators (e.g., Lancet et al. 1980) using radiolabelled 2-deoxyglucose that the intensity of metabolic activity is nonuniform through the bulb. The locations, sizes, and shapes of foci of relatively high rates of cellular oxidation were similar to those of the foci detected electrophysiologically in the bulb. However, the conclusion differed: the radiolabelling method was restricted to one odor for each subject, so that dependence of activity pattern on subject, odor, and expectation could not be sorted out.

The hypothesis was proposed that behavioral conditioning caused an increase in the strength of selected bulbar synapses. The first candidate considered was the synapse between the receptor axon and the mitral cell. This was quickly disposed of by delivering pulses antidromically to the bulb by electrical stimulation of the olfactory tract. The amplitude of evoked oscillation conformed to the spatial pattern of burst amplitude, and it changed in the same way with training, even though the test input did not pass through the designated synapse (Freeman and Schneider 1982). The same test eliminated the external interneurons, which were found not to respond to the electrical stimulus in the tract (Freeman 1975).

The mutually excitatory synapses (k_{cc}) were implicated by an earlier experiment in which cats were trained to press a bar for milk when an electrical stimulus was delivered to the tract. The evoked potential in the cortex (the dynamics of which is also represented by a KII set) was found to change in a characteristic way as the animal learned to attend the stimulus. The same pattern was found in the bulbar antidromic response in rats. The KII set was used to simulate these impulse responses and their changes by representing the input with a Dirac delta function. The only change in coefficient that was necessary and sufficient to reproduce the change in evoked potential with learning was a small increase in k_{ee}. Further simulation of the KII model with simulated receptor input showed that the k_{ee} coefficients for local neighborhoods were critically important for determining the threshold for the onset of limit cycle activity. For example, an increase in k_{ee} on the order of 40% could increase the sensitivity for induction of the limit cycle by a factor of 40,000. When the values for k_{ee} were uniform across the surface of an array of KII subsets, the spatial pattern of simulated bursts conformed to the pattern of input. When the k_{ee} values were unequal, the spatial pattern of burst amplitude conformed to the spatial pattern of k_{ee} and not to the input (Freeman 1975, 1979b,c).

Cooperative Processes of Abstraction and Generalization

This result led to the hypothesis of a kind first proposed by Hebb (1979) that during training to respond to an odor, a template of connections is strengthened among those mitral cells that receive input simultaneously on each of a succession of inspirations, provided that the reinforcing stimulus activates an enabling centrifugal pathway, such as an aminergic tract from the brainstem. This view is based on the assumption that among the 10^8 receptors in the nose, only a fraction (such as 10^6 to 10^7) are sensitive to any one test odor, and that during any one inspiration, only a subfraction (such as 10^2 to 10^5) receives the odor. Those that do receive input then coactivate mitral cells, and their mutually excitatory synapses are conceived to be irreversibly modified if there is reinforcement. Turbulence in the air flow can be assumed to deliver the odor to a different subset of receptors on each inspiration, so that, over the several hundred odor-bearing inspirations in a training session, the cumulative subset of activated receptors and mitral cells may approach the total set that is capable of activation.

Once such a template is laid down, an interconnected nerve cell assembly is formed. The KII simulation predicts that if any subset within the set of sensitive receptors stimulated during training receives input, the entire assembly of mitral cells interconnected by the template is activated in a stereotypic manner, whether or not that particular receptor subset received simultaneous input during the training period (Figure 9.10). The selective increase in sensitivity to the test odor implies that the assembly acts to abstract the test odor from the background; the stabilized pattern of activity suggests that the assembly serves to generalize over the set of receptors that are equivalent in respect to the test odor. If, as the experiments show, the output of the bulb is dominated by the pattern of the oscillatory burst, then the information received by the cortex is dominated by the template pattern and not by the input, although other input is not thereby excluded.

In brief, we can say that the inner bulb processes information by generating a carrier wave (Figure 9.11). The preprocessed sensory image in the outer layer of the bulb provides the patterned input, and the bias excitation acts to induce a state change from equilibrium to a limit cycle. Thereafter, the event develops with spatial amplitude modulation of the carrier wave that depends on the chemical state of the bulb under centrifugal controls reflecting arousal, and on past experience embedded in altered synaptic connections among projection neurons. The process results in abstraction of selected aspects of input and in generalization into selected classes of receptor input that represent familiar categories. The bulbar output undergoes a spatial integral transformation in the tract, and the resultant establishes a new spatial image in the olfactory cortex, subject also to synaptic modification by learning (Freeman 1975). These several neural processes can easily be recognized as the main elements of the psychological process of perception (Grossberg 1980, Freeman 1981).

Limitations of time and space will not permit me to consider other aspects of information processing. These include the response of this system to unexpected input and the induction of the orienting response; its response to unwanted input by processes of adaptation and habituation; the formation and testing of multiple and compound templates for different odors; the flexible and periodic shifting from one template to others; the selective recall of templates; the formation by the cortex of motor templates and images; and other equally fascinating features of this primitive cognitive processor.

Conclusion

My intent has been to describe in a nonmathematical way certain biophysical properties of large ensembles of nerve cells. The experimental work must be done in a context of goal-oriented behavior in order that the relevant properties be manifest. My key point is that the cerebral cortex cannot be effectively portrayed as a network of single neurons analogous to transistors. There must exist entities that transcend the neuron by virtue of widespread synaptic action among neurons. The assembled evidence indicates that macroscopic, cooperative activity does exist in the brain, that it is mediated by synaptic transmission and not by chemical diffusion or electrical fields, and that it participates in operations performed on sensory input to the brain. It is analogous to the property of temperature reflecting the kinetic energy of molecules in a gas. As temperature is defined for the ensemble and not the particle, so sensory information is defined for the neural ensemble and not for the neuron. Like temperature, the interactive condition can be regarded as an operator. Once the limit cycle activity is induced, it destabilizes millions of neurons, drives them into coherent rhythmic activity, and directs their organized output onto widely disseminated target neurons. It generates a new image that incorporates an input image with preexisting patterns that are collated residues of previous images and with more general factors relating to the states of the brain and body. In a process of self-organization, it creates an image of a sensory event in a context of past experience and present expectation. This outcome satisfies a fundamental postulate in neuroscience, which holds that psychodynamics is an expression of neurodynamics; ultimately, they must conform. The work I have sketched shows that neurodynamics is a member of the family of dynamics, yet it differs from its siblings in its substrate and postulates as much as the others—such as thermodynamics and hydrodynamics—differ from each other. This is primarily because the state variables do not represent concentrations of matter and energy directly, but indirectly via their expressions in dendritic and axonal activity patterns.

Addendum

Three advances have been made in the three years since this chapter was first written. First, rabbits were appetitively conditioned to discriminate and respond differentially to two odors (Viana Di Prisco and Freeman 1985). Second, new techniques were devised and developed to measure the EEG more precisely (Freeman 1987). The measurements in those animals have shown that, after conditioning, a stable spatial pattern of neural activity appears in the bulb that is: (1) unique to each odor (including the background complex), (2) present only when the conditioning odor is present, and (3) changes in form when the reinforcement contingencies are modified or a new conditioning odor is introduced. Third, the basal activity state is not an equilibrium under noises; it is chaotic (Skarda and Freeman 1987).

Our interpretation is that the formation of a nerve cell assembly during learning proceeds in the manner described in the chapter, and that this assembly serves to provide the structure of a learned limit cycle attractor in the bulbar mechanism (Freeman and Viana Di Prisco 1986b; Freeman and Baird 1987; Freeman and Grajski 1987). We believe that an attractor forms for each odor that an animal learns to discriminate. Its basin is determined by the receptors for the odor that were activated during learning. The attractor determines a spatial pattern of amplitude of neural oscillation over the entire bulb that is unique to it, and that gives a unique character to the bulbar output for that odor (Freeman and Viana Di Prisco 1986a).

These new findings validate the earlier experimental findings and computations, and they further open the brain to exploration in the context of nonlinear dynamics (Freeman 1987). What they disprove and invalidate is the concept that the bulb forms "images" of expected sensory input that can serve as "representations" of the outside world. We believe now that the notion of representation in all of its many forms (e.g., images, codes, symbols, tables, engrams) is not only inconsistent with physiological data, but is unnecessary to explain animal behavior in terms of brain mechanisms (Freeman and Skarda 1985; Skarda 1986). Animals and people are engaged interactively with their environments, but they need not—and, we believe, do not—construct "world models" as their bases for coping. The brain is a self-organizing process and structure that incorporates, adapts to, and modifies the environment through its control of behavior. The implications of this new view of psychology, philosophy, and artificial intelligence are discussed elsewhere (Skarda 1986; Skarda and Freeman 1987).

Bibliography

Babloyantz, A., and L.K. Kaczmarek, 1979: Self-organization in biological systems with multiple cellular contacts, *Bulletin of Mathematical Biology* 41:193–201.

Bressler, S.L., 1982: *Spatiotemporal analysis of olfactory signal processing with behavioral conditioning*, Ph.D. thesis in Physiology, University of California at Berkeley.

Bressler, S.L., and W.J. Freeman, 1980: Frequency analysis of olfactory system EEG in cat, rabbit, and rat, *Electroencephalography of Clinical Neurophysiology* 50:19–24.

Freeman, W.J., 1975: *Mass Action in the Nervous System*, Academic Press, New York.

Freeman, W.J., 1978: Spatial properties of an EEG event in the olfactory bulb and cortex, *Electroencephalography of Clinical Neurophysiology* 44:586–605.

Freeman, W.J., 1979a: Nonlinear gain mediating cortical stimulus-response relations, *Biological Cybernetics* 33:237–247.

Freeman, W.J., 1979b: Nonlinear dynamics of paleocortex manifested in the EEG, *Biological Cybernetics* 35:21–37.

Freeman, W.J., 1979c: EEG analysis gives model of neuronal template-matching mechanisms for sensory search with olfactory bulb, *Biological Cybernetics* 35:221–234.

Freeman, W.J., 1980: Use of spatial deconvolution to compensate for distortion of EEG by volume conduction, *IEEE Transactions on Biomedical Engineering* 27:421-429.

Freeman, W.J., 1981: A physiological hypothesis of perception, *Perspectives on Biological Medicine* 24:561-592.

Freeman, W.J., 1987: Techniques used in the search for the physiological basis for the EEG, in A. Gevins and A. Remond (eds.), *Handbook of Electroencephalography and Clinical Neurophysiology,* Vol. 3A, Part 2, Chapter 18, Elsevier, Amsterdam, Holland.

Freeman, W.J., and B. Baird, 1987: Relation of olfactory EEG to behavior—spatial analysis, *Behavioral Neuroscience* (in press).

Freeman, W.J., and G. Grajski, 1987: Relation of olfactory EEG to behavior: factor analysis, *Behavioral Neuroscience* (in press).

Freeman, W.J., and W. Schneider, 1982: Changes in spatial patterns of rabbit olfactory EEG with conditioning to odors, *Psychophysiology* 19:45-56.

Freeman, W.J., and C.A. Skarda, 1985: Spatial EEG analysis, nonlinear dynamics, and perception—the neo-Sherringtonian view, *Brain Research Reviews* 10:147-175.

Freeman, W.J., and G. Viana Di Prisco, 1986a: EEG spatial pattern differences with discriminated odors manifest chaotic and limit cycle attractors in olfactory bulb of rabbits, *Proceedings, Conference on Brain Theory* (Trieste, Italy, 1984), 97-119, Springer-Verlag, Berlin.

Freeman, W.J., and G. Viana Di Prisco, 1986b: Relation of olfactory EEG to behavior—time series analysis, *Behavioral Neuroscience* 100:753-763.

Grossberg, S., 1980: How does a brain build a cognitive code? *Psychological Review* 87:1-51.

Hebb, D.O., 1979: *Organization of Behavior,* John Wiley, New York.

Lancet, D., C.A. Greer, J.S. Kauer, and G.M. Shepherd, 1982: Mapping of odor-related neuronal activity in the olfactory bulb by high-resolution 2-deoxyglucose autoradiography, *Proceedings of the National Academy of Sciences USA* 79:670-674.

Skarda, C.A., 1986: Explaining behavior—bringing the brain back in, *Inquiry* 29.

Skarda, C.A., and W.J. Freeman, 1987: Brains make chaos to make sense of the world, *Brain and Behavioral Science,* July.

Viana Di Prisco, G., and Freeman, W.J., 1985: Odor-related bulbar EEG spatial pattern analysis during appetite conditioning in rabbits, *Behavioral Neuroscience* 99:964-978.

Neural State Variables

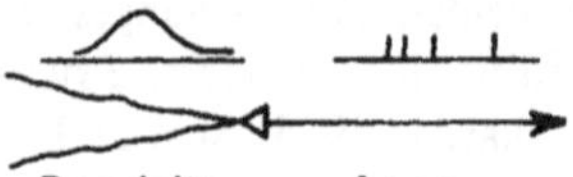

Form	Dendrite	Axon	Scales	
			Time	Distance
Structure	Dendrite	Axon	Time	Distance
Neuron:				
activity	membrane polarization	action potential frequency	msec	μm (point)
observable	postsynaptic potential	unit train		
Ensemble:				
activity	current density	pulse density	sec	mm (area)
observable	EEG wave	unit sum		

FIGURE 9.1 Neuro-electrical activity occurs in two modes, the impulse action characteristic of the axon and the wave action characteristic of dendrites. The observable manifestations are not identical to the active state; appropriate transforms must be constructed to derive estimates of state variables from electrical recordings. The state variables of single neurons and of neural ensembles are defined and evaluated in different ways.

Conversion Operations

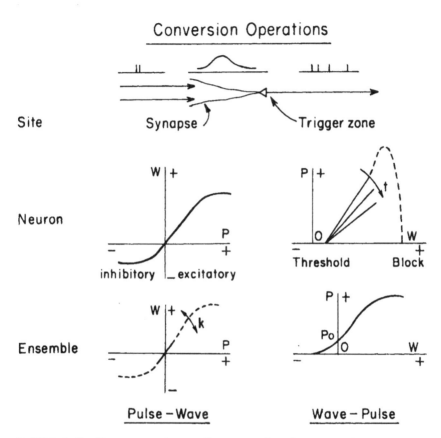

FIGURE 9.2 The operations of conversion between the wave W and pulse P modes of activity take different forms for the single neuron and the ensemble. At the microscopic level P to W is time-varying and nonlinear, and W to P is time-varying (t) but linear between threshold and cathodal block. At the macroscopic level W to P is static and nonlinear, whereas P to W is constrained to s small-signal linear range. The slope dW/dP changes with learning; the slope dP/dW changes with arousal.

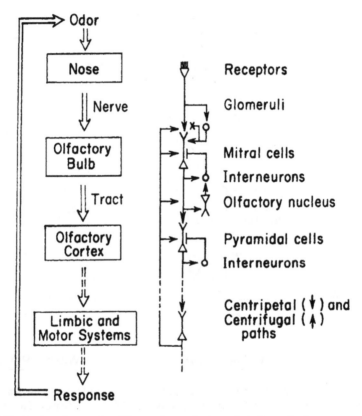

FIGURE 9.3 The main flow of activity in the olfactory system is forward (centripetal) through three stages. Feedback
occurs at all levels within, between, and around the stages.
Feedback from the brain into the bulb and cortex is denoted
centrifugal. There are 10 to 12 centrifugal paths with
various functions.

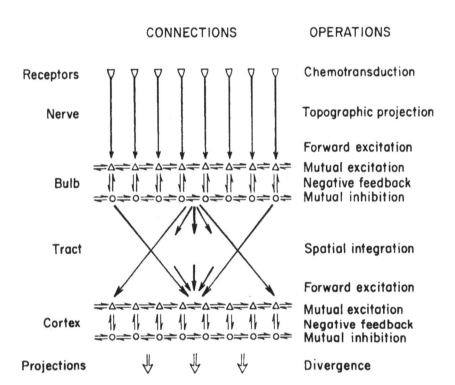

CONNECTIONS   OPERATIONS

Receptors   Chemotransduction

Nerve   Topographic projection

          Forward excitation
          Mutual excitation
Bulb   Negative feedback
          Mutual inhibition

Tract   Spatial integration

          Forward excitation
          Mutual excitation
Cortex   Negative feedback
          Mutual inhibition

Projections   Divergence

FIGURE 9.4   Each layer is organized into a sheet of neurons.
The state variables are defined for the pulse and wave
modes in the two surface dimensions. They are discretized
at intervals corresponding to the spatial coarse-graining of
the layers.   Interactions occur laterally in each layer.
The nerve provides for topographic projection, whereas the
tract provides for spatial integration.

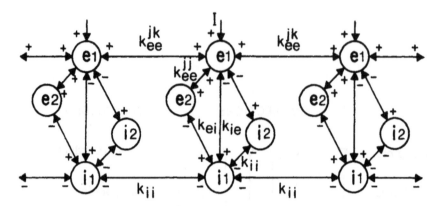

FIGURE 9.5 The topology is shown for three representative
KII subsets embedded in a KII set. Each KI set is represented by two KO sets in a feedback relation; this seems to
express the ensemble property that at any moment only a
small fraction of component neurons are either transmitting
or receiving pulses. The superscripted coupling coefficients
k are subject to modification by learning. e, excitation;
i, inhibition. From Freeman (1979c).

$$F(v_n) \triangleq \frac{1}{ab} \frac{d^2}{dt^2} \left[v_n(t) \right] + \frac{a+b}{ab} \frac{d}{dt} \left[v_n(t) \right] + v_n(t)$$

$$F(v_{e1,j}) = \zeta_e^j k_{ee}^{jj} p_{e2,j} - \zeta_e^j k_{ie} (p_{e1,j} + p_{i2,j}) + \sum_{k \neq j}^{N} \zeta_e^j k_{ee}^{jk} p_{e1,k} + I_j$$

$$F(v_{e2,j}) = \zeta_e^j k_{ee}^{jj} p_{e1,j} - \zeta_i k_{ie} p_{i1,j}$$

$$F(v_{i2,j}) = \zeta_e^j k_{ei} p_{e1,j} - \zeta_i k_{ii} p_{i1,j}$$

$$F(v_{i1,j}) = \zeta_e^j k_{ei} (p_{e1,j} + p_{e2,j}) - \zeta_i k_{ii} p_{i2,j} - \zeta_i k_{ii} \sum_{k \neq j}^{N} p_{i1,k}$$

FIGURE 9.6 The differential equations are summarized for
the KII set. The linear operator must be second-order or
higher for neural as distinct from psycho-physical modelling.
v, wave density; p, pulse density (see FIGURE 9.7); I, input
density (a time function). The derivation is in Freeman
(1975, 1979b,c).

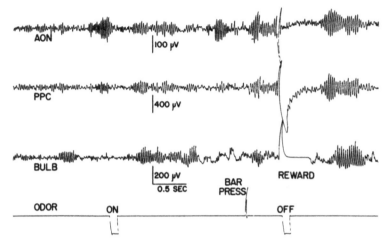

FIGURE 9.7 An example is shown of the EEG activity from the olfactory bulb, anterior olfactory nucleus (AON), and olfactory (prepyriform) cortex (PPC) of a subject with implanted electrodes. At ON a valve is opened to inject an odor into an on-going air stream; it arrives 0.5 second later. The subject presses a bar 1.5 seconds thereafter to receive water.

$$Q = Q_m \left\{ 1 - \exp\left[-(e^v - 1)/Q_m \right] \right\}, \ v > -\mu_o$$

$$Q = -1, \ v \leqslant \mu_o$$

$$\mu_o = -\ln\left[1 - Q_m \ln (1 + 1/Q_m) \right]$$

$$p = \mu_o (Q + 1)$$

$$\frac{dp}{dv} = \mu_o \exp\left[v - (e^v - 1)/Q_m \right]$$

$$v_{max} = \ln (Q_m)$$

FIGURE 9.8 The equations are summarized for the nonlinear conversion of wave density u to pulse density p. Q and v are normalized pulse and wave densities to v=Q=o at rest. The positive exponential term e^v is an expression at the level of the ensemble of the sodium activation factor in the Hodgkins-Huxley system. The maximal ensemble firing rate Q_m (relating also to u_o and p_o) is determined by multiple factors (such as the early potassium current), reflecting ultimately the ergodic property of the ensemble over the pulse-recovery cycle for single neurons. From Freeman (1979).

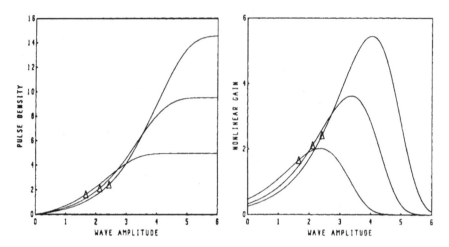

FIGURE 9.9 In the right frame are shown three examples of
a curve fitted to statistical data showing W to P conversion.
In the left frame are shown the derivatives dP/dW of the
curves. The triangles represent resting or equilibrium
values. With increasing wave amplitude there is a coupled
increase in pulse density and in gain dP/dW. From Freeman
(1979a).

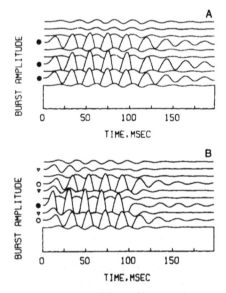

FIGURE 9.10 A. The output is shown for a KII set with
input to 3 elements. The coupling coefficients for mutual
excitation k_{ee} are increased thereafter.
B. The model that now has a template corresponding to
the "learned" input is given input to one of the 3 elements.
The pattern of output corresponds to the template. From
Freeman (1979c).

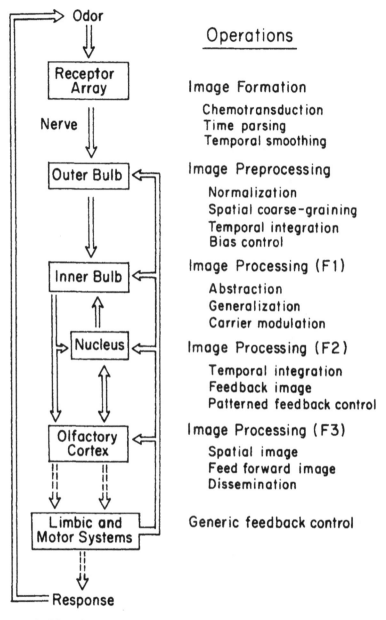

FIGURE 9.11 The principal operations of information processing in the olfactory systems are summarized in so far as they are understood and at the stages where they occur. The designations F1 and F2 denote the possible locations of functional layers corresponding to the system proposed by Grossberg (1980). Note that image preprocessing is completed prior to the suppression of topographically specific information.

10. Prolegomena to Evolutionary Programming

Turing and Darwin

Computer science as an academic discipline had its origin in the invention of the electronic digital computing machine. Conceivably, a computer science could have developed independent of any machine at all. For thousands of years, people studied computation processes. Some simple computing devices, such as the abacus, were invented. But by and large, it was humans who did the computing. Formalisms for describing computing processes were developed to some extent independently of the machine, though I am sure historians of science could document a common intellectual undercurrent. Turing was involved with machines, but his conceptually transparent Turing machine is a formalism for computing, not a real machine at all. The Turing formalism consisted of a finite-state abstraction of an organism (a finite automaton) along with a tape that it can move and mark. Turing thought of the states of the finite automaton as mental states, say of a child doing arithmetic on a notepad (represented by the memory tape). Under this *interpretative postulate*, the Turing formalization established a plausible link between algorithmic and psychological processes that was to play an important role in the cognitive sciences.

It turned out that Turing machines defined the same class of functions as a large variety of alternative formalizations of algorithmic processes. This equivalence led Turing and others to assert that any effectively computable function could be computed by a digital process.

This claim, known as the Turing-Church thesis, is unprovable but useful. Possibly its authors were thinking only of a particular class of models of computation, such as recursive functions, production rules, and other discrete symbol manipulation processes. But one can always propose alternative models of computation, such as analog processes. Thus, the scope of Turing-Church inevitably expanded to the claim that digital computation processes can be used to simulate any physically possible process in nature; if it cannot be so expanded, the unsimulatable process could be used to effectively solve problems that cannot be solved by a digital process (Conrad and Rosenthal 1980). At the very least, one would have to admit that the unsimulatable process is refractory to mathematical description, or that some forms of mathematical description are refractory to reformulation in terms of Turing-type formalizations of computation processes.

Having so expanded the concept of computation, one is forced to admit that all the information-processing activities of organisms can be viewed as forms of computation. Pattern and object recognition are as legitimately forms of computation as is arithmetic. This is not surprising today, since pattern and object recognition is one of the chief things for which we try to use computers. But once this is admitted, we then also have to admit that computation processes have an existence that is prior not only to the electronic digital computing machine but to humans as well. Unless we define computation processes in terms of a limited class of preferred formalisms, it is necessary to go still further and admit that any process in nature can be viewed as defining some type of computation process. Whether such processes are usefully viewed as contributing to problem solving is another question. A digital computer might spin along without solving any problem from our point of view, yet we would be hard pressed to say that it is not computing. To avoid this difficulty, we will say that a system does useful computing if it appears to solve problems. Of course, every system solves the problem of being itself.

Having gone this far, we can embolden ourselves to take a step back—back to Darwin and Wallace. They proposed that evolution by variation and selection is the fundamental mechanism of problem solving in nature. The evolution method is fundamental, since it is not parasitic on an outside intelligence. No programmer is needed. It is self-programming. According to Turing-Church, it should be possible to simulate evolution. There is no absolute distinction between evolutionary problem solving and algorithmic problem solving. We can view both in terms of models of computation and ask how much equivalent Turing machine computation they involve, assuming that we take Turing machines as our reference machines. We could imagine a machine quite suitable for evolution and take this as the reference machine. Perhaps an arithmetic process would appear to require an enormous amount of evolutionary computation on the part of such an evolutionary reference machine. The issues of complexity take on quite a different aspect when looked at in this light.

As it happens, this is not the way computer science developed. The invention of the actual machine—a history-altering event in the general sense—played the main role. Before the advent of the digital computer, it was humans that were viewed as computers, not machines or other organisms. After the development of the digital computer, the term "computer" was used to refer to machines (Burks and Burks 1981). Whether humans and organisms are computers that work like computing machines became the controversial question. McCulloch and Pitts (1943) made a big impact on this issue by showing that it is possible to realize any finite automaton by wiring up threshold elements that look like all-or-none neurons according to simple rules. This added a dramatic neuroanatomical dimension to Turing's link between algorithmic and psychological processes. A connection to logic was also established. Logic gates could be represented by particular formal neurons or simple combinations of formal neurons. Memory could be represented by delay loops in the network structure. For the first time, it was possible to understand with Cartesian clarity how complex behavior might be embodied in a natural system.

Many neuroscientists now disparage the simplicity of the McCulloch-Pitts model, and many computer scientists dismiss its engineering significance. But I believe that the model has provided the underlying presupposition of these fields. What McCulloch-Pitts made clear was that it is possible to synthesize complex machines with desired behavior from a few simple, easily made devices; and that it is possible to decompose complex machines with elaborate behavior into networks of such devices in a definite way. I will call such systems *structurally programmable,* since it is possible to control their behavior by manipulating the switches and connections according to well-defined rules. The specialization of computer science into software and hardware branches is based on the assumption of structural programmability. Neuroscientists and engineers may claim that they reject what McCulloch called "the poverty stricken simplicity of threshold-like neurons." Most, however, accept the principle of structural programmability in their efforts to design machines and languages or to analyze neural systems. Later, I shall argue that structural programmability is not a necessary feature and that the link between Turing and Darwin can only be made through structurally nonprogrammable systems.

Impact of the von Neumann Machine

To understand the development and current state of computer science, it is of course necessary to consider actual machines. The history of the machine is complicated, but for our purposes, we can take

ENIAC as the first general-purpose digital computer and place it in the period 1940–45 (see Burks and Burks 1981). The invention of the machine created a sequential interplay of problems and solutions and also a sequential interplay of these with perceived possibilities for application. I have listed the main problems in Table 10.1. The first was setting the switches and connecting the wires of the machine to perform a desired task. The stored program machine required only switch setting, and the first problem was to express these settings in codes that could be communicated to the machine. These languages were hard to work with, and so a prime problem immediately became that of developing high-level languages. These are simply formalizations of algorithmic processes that are more convenient from the point of view of human thought processes. The next problem became that of linking programs written in high-level languages to the machine language, that is, to the state of the machine. This is the problem of compiler construction.

The existence of high-level algorithmic languages, such as Algol, Fortran, and LISP, increased the possibility for applications. The productivity of programmers was increased at least ten-fold; but even more important, high-level languages made it possible for the ordinary person to use the computer with a minimal amount of training. So the number of people who wanted to use the machine naturally increased, inevitably leading to the problem of managing usage. This is the problem of operating systems. It soon became clear that the most severe bottleneck was the slow rate of input and output. The central computer was sitting idle while the peripherals were overloaded. The development of time-sharing systems thus became the next problem in the sequence.

The first computers were made out of unreliable vacuum tubes, and hence the problem of reliability also loomed as a big initial issue. Some clever designs were developed that allowed reliable computation with unreliable components (von Neumann 1956). These designs utilized redundancy and were reminiscent of nervous system organization. But the development of much more reliable solid-state components provided a physical solution to the problem of reliability, at least for a while. The development of integrated circuits enormously increased computing power and very large computations inevitably eventually incur errors. Thus, the problem of tolerance to error (fault tolerance) has reappeared as an issue.

In the early days of computing, efficiency was an issue. Programmers considered it a virtue to put tricks into their code that would speed up execution. The development of integrated-circuit technology enormously increased computing resources. It became evident that the human effort involved in writing, correcting, reading, and maintaining code had by and large become more costly than running the machine. Programming style became the issue, and the concept of structured programming came to the fore. Many investigators believe that the next logical step on the programming side is to develop specification languages, that is, languages that specify goals at a very high level. But it appears that it is very hard to do this is a way that is unambiguous and at the same time more humanly understandable than our present high-level programming languages. Another issue is that of program flexibility and adaptability. Structured programs are clearly easier to work with than highly intertwined programs. But, unfortunately, programs are inherently fragile. It is possible to build program architectures with a high degree of modularity and modifiability. But to a large extent, this modifiability resides in the mind of the the programmer. It is difficult to document this mental image and to communicate it to other programmers, or even to communicate it to oneself at a later point in time. The art of good documentation would thus appear to be an issue that would naturally emerge and re-emerge in the development of computer science. But the impression of this author is that due to some sociological or psychological constraint, it has never really been taken seriously.

The original trend in computing was to bigger and bigger machines housed in large, centralized computer centers. Increases in speed and decreases in size associated with VLSI technology made the small computer into a powerful tool. The proliferation of small computers brought the problem of communication among separated machines to the fore. Networking and distributed computing became major issues. Some investigators believe that distributed computing can serve to increase the efficiency with which resources are used in parallel. Whether this is the case is open to serious question due to the number of resources that must be devoted to messengers, protocols, and guards. However, it is undoubtedly the case that the increase in computing resources has made parallelism into the most pressing problem of today's computer science. The paradox of increased computing resources is that it has brought decreased efficiency in the use of these resources. This is due to the fact that sequentiality is necessary for effective programmability. But sequentiality means that most of the resources of the machine are dormant at any given time. As the number of resources increases, the fraction that are active becomes smaller.

Parallel and multiprocessor architectures can be built and are being built. The problem is that traditional programmability breaks down. If a machine executes two or more processes at once, it is always possible that a conflict will occur. If conflicts are precluded, it is likely that most of the processors will lie idle while the processor with the longest task works away. A great deal of effort is going into research on parallel compilers and into languages with parallel points of view, but it is clear that the problem of efficient use of resources is inherently difficult. On the theoretical side, these efforts are mirrored by the enormous amount of work that has been done by theorists on the resource demands of different algorithms. Some problems blow up exponentially, and it is clear that no success in speedup or parallelism will help significantly with these (apart from approximate solutions). To the extent that a problem can be put into a form in which the resources required grow as a polynomial function of problem size it is potentially possible that effective use of parallelism will have a big impact, assuming that an algorithm can be found that admits some kind of parallel decomposition.

AI and the Redefinition of Intelligence

Some of the developments in computer science, such as data base design, have been driven principally by demands on the applications side. Artificial intelligence has been driven both by the applications side and by the desire to understand intelligence. Intelligence here can be viewed as a natural thing, or as something completely de novo in the universe. If AI is defined as the attempt to make digital computers intelligent and intelligence is viewed as correlating to human intelligence, it is necessary to admit a very big assumption. The structure of the brain is quite unlike that of a digital computer. If intelligence could be understood purely in terms of the algorithms that can be executed by present day machines, this would mean that the material organization of the brain is an irrelevancy so far as pattern recognition, object identification, problem solving, learning, and language acts are concerned.

The hypothesis here should be stated more accurately. Much of the effort in AI has involved communicating a representation of the external world to the machine. For example, the rules in an expert system are such a representation. The assumption is that available computing power is sufficient to fit an adequate representation into the machine. The second assumption is that nothing significant is lost by virtue of the fact that the representation of reality in the computer is a coded reality and not the real reality.

AI has had two major successes. The first is in clarifying and, indeed, redefining the nature of intelligence. This redefinition is graphically illustrated by the story of Clever Hans, a horse that stunned the world more than half a century ago by doing arithmetic. Clever

Hans, it turned out, was watching subtle, unconscious motions of his master that signaled the answer. People at the time decided that the horse was not so clever. From the standpoint of what machines can do, we now realize that arithmetic does not involve very much intelligence. Pattern recognition and learning of the type Clever Hans had to do to pick up the signals is hard. AI has served to clarify and redefine what is easy and what is difficult so far as intelligence is concerned.

The second success of AI is in the domain of specific applications of sophisticated programming techniques to problems in constrained universes, such as the constrained universes of games, block worlds, and to the recognition of complex but unambiguous patterns (such as universal product codes). As soon as ambiguity is admitted into the universe, AI techniques are stopped short by the combinatorial explosion. Tasks easy for Clever Hans, such as pattern and object recognition, become totally refractory for computers. Presumably, combinatorially explosive tasks are also refractory for biological organisms. But the enormous parallelism of organisms and the fact that they are the products of evolution rather than of programming allows them to deal with very much larger problems in an approximate way, assuming that an approximate solution with polynomial time character exists. The difficulty of dealing with ambiguity suggests the importance of learning, evolution, and self-organization. These were ideas that were prominently considered at the beginning of the AI enterprise, but were given up in favor of high level algorithmic approaches. These approaches have now run into problems that bring the issue of learning back to the fore.

Because of these problems, some segments of the computer science community have in recent years become interested in new architectures. There is a recognition that the material organization of the machine is important. Some of these architectures have a brainlike connectionist aspect. The similarity to the brain is that the convergence and divergence of inputs to the basic components are high, just as the divergence and convergence of axons and dendrites to and from neurons in the brain is high. In reality, these models are quite unbrainlike in that the components are simple switching elements. The neurons in the brain are themselves complex organisms with an enormous number of internal molecular switching processes. Nevertheless, connectionist models with simple switches do offer new possibilities for parallel computation. Cellular automata—that is, simple machines with neighbor-neighbor interactions—provide another model for parallel computing, and some new architectures are built around this idea. Cellular automata are close to physical models in that partial differential equations are represented in the computer as cellular automata in which the state set of each cell runs over the real numbers. Again, all these ideas were prominent early in the history of computer science, but slipped into the background due to the faster progress that could be made with the easily controllable von Neumann architecture.

Prior to the advent of chip technology, many scientists interested in modeling natural processes by means of differential equations considered analog computers to be the vehicle of choice. The same sharp increase in computer resources that led to the development of structured programming, to the distribution of computing, and to the paradox of parallelism led most such modelers to shift to digital computers, and where possible, to vector supermachines. Anyone who models physical or biological systems soon discovers that the number of resources in nature and their parallelism vastly exceed the resources of any programmable machine. The number of neurons, the number of chemical processes in a neuron, and the number of particles in a chemical process that a computer can cope with are minute by comparison to the number that actually occur. This limitation of computer modeling could have been taken as an omen. If the material organization of the brain is important for human intelligence, it is going to be very difficult to duplicate human intelligence in the material organization of a digital computer. On the more positive side, it might be taken as a hint that new architectures are needed and that these architectures might usefully incorporate analog and other dynamic physical processes.

A Residue of Difficult Problems

Computer science made dramatic progress with early problems such as languages and compilers. These were difficult, but solutions existed and capable individuals and groups worked them out. Operating systems present some inherent difficulties, but computer scientists have developed models for designing them in a manner that gets the job done. Progress with problems of greatest current concern—specification languages and adaptable code, parallelism, program correctness, AI in natural environments, and computer learning—has been much slower. Perfect solutions cannot be expected. These are a residue of inherently difficult problems. Slow progress can probably continue, but dramatic advances will probably have to await new points of view about the character of the solutions and the methods of obtaining them. Yet, these problems cannot be ignored. The increasing importance of computers in industry, finance, medicine, and warfare has made these into societal problems. If computer programs are delicate, it is humans that have to become adaptable enough to protect them from disturbance. If AI requires highly constrained universes, it is humans and human institutions that will have to conform to rigid rules of behavior. The impending paradox is clear. The integration of computers into human society demands that humans be more adaptable on the one hand, and more rigid on the other.

Let us imagine that evolutionary processes could be efficiently implemented in computers. Biological organisms are an existence proof for the effectiveness of evolutionary problem solving in nature. Over the course of evolution species learn to solve problems by the self-programming process of variation and selection. Table 10.2 lists the attributes of evolutionary programming. Each of these corresponds to one of the "tough nuts" in Table 10.1. The evolutionary programmer specifies goals and criteria, rather than the rules that constrain the machine to meet these goals and criteria. A specification language is the natural language tool. Fault tolerance and program correctness are viewed in a different light, since error is a prerequisite for evolution. Evolution by variation and selection is a fundamental form of adaptability. If a system can learn through evolution, it can learn to deal with a variety of environments. The degree of constraint required in the universe, the main limitation of conventional AI, is decreased. Since prescriptive programming is not possible, sequential operation of the machine becomes entirely unnecessary. The breakdown of conventional programmability—of the user's prescriptive control over the machine—is no longer a crisis that must be contained. Evolution is a population phenomenon, and as such, it is advantageous for many variants of a process to run in parallel. Highly parallel machines are the natural hosts for evolutionary computations. Furthermore, evolutionary systems can learn to use their resources efficiently in parallel if this is specified as one of the criteria on which selection is based.

What our discussion shows is that biological systems have managed to deal with the chief problems that face today's computer science. If computers could support evolution, it would be possible to manage these problems in our technical systems. Unfortunately, computers so far have not been good evolvers. The idea of using variation and selection mechanisms has been attempted at various stages in the development of computer science. For example, Fogel et al. (1966) wrote an early book that applied variation and selection to the evolution of finite automata. The methods worked, though the task was not hard. Bremermann (1962) achieved success with an evolutionary optimizer applied to unimodal optimization on a multidimensional surface (see also Rechenberg 1973). Holland (1975) developed a general framework, called a genetic plans framework, which has been applied to production-like systems. Our group has worked on models of evolutionary process in nature (e.g., Conrad 1981; Conrad and Strizich 1985; Rizki and Conrad 1985) and on neuronal systems that learn to recognize patterns or navigate towards targets through an evolutionary algorithm (e.g., Conrad 1981; Kampfner and Conrad 1983; Kirby and Conrad 1985). But bold efforts to evolve computer code or attack high-level problems have been chilling failures.

The source of this failure is the delicacy of computer code. A single change usually renders the code unacceptable. Minsky (1961) called this the mesa phenomenon, referring to the fact that peaks are separated by untraversable flatlands. The term "crevice phenomenon" is probably more descriptive. A single alteration in a functional program causes it to fall into a deep adaptive crevice. It is clear that if improvements require more than a single structural change in the program (mutation of elements of code, recombination or crossing over of blocks of code), the rate of evolution will scale as p^n, where p is the probability of the alteration and n is the number of alterations that must occur simultaneously. The rate of evolution will clearly become unacceptably small for $n > 1$ (see Conrad 1972, 1983).

We have now arrived at a contradiction. According to Turing-Church, it should be possible to model any process in nature with digital computers, provided that the time and processor resources available to the computer are adequate. Evolution is a process that occurs in nature. Thus, it should be possible to model the evolutionary process, as claimed in our earlier juxtaposition of Turing and Darwin. If we can model the evolution process with digital computers, this means that it should be possible to manage the tough nut problems of computer science in the fashion of biology systems. I use the term "manage," since the important point for organisms and societies is achieving good enough solutions to compete effectively in the game of life.

Towards Nonprogrammability

To understand the reason for the gap between theory and practice, we have to return to the conceptual roots of computer science. We can recall that computer science early on divided itself into software branches and hardware branches. Automata theory—the abstract theory of computing—divided itself into two corresponding branches. The first, behavior theory, dealt with the capabilities of machines, their state-to-state behavior, and the languages (or sequences of input symbols) that could set them to a particular state. Behavior theory is thus closely related to languages and compilers. The second branch was structure theory, which dealt with synthesis from or decomposition into elementary machines, and was closely related to hardware issues. The mathematical theories of these two different branches were largely separate. It was recognized that in many cases, one might not be able to find a particularly efficient synthesis or decomposition. For example, a definite procedure exists for constructing a McCulloch-Pitts network to realize any finite automaton; but it is quite possible that this procedure will lead to a network with an unreasonably large number of formal neurons. But aside from this efficiency issue, the prevailing view was that the different realizations of a network were all equivalent from the standpoint of languages and algorithms. This fit into the AI assumption that the processes of intelligence could be divorced from the material organization of the system.

This divorce becomes much less tenable when evolutionary processes are admitted. This is because evolutionary processes involve random changes in the structure of a system. Either the elemental machines must change their character or the connections among these machines change. Imagine two different machine structures, A and B, that are completely equivalent as far as their state-to-state behavior is concerned. By equivalent we mean that A simulates B and B simulates A. The languages, or sequences of symbols, that the two machines recognize or generate are completely equivalent. However, this does not mean that a random structural change in A will have the same effect as a random structural change in B. A random structural change in A will produce a de novo automaton A′ and a random structural change in B will produce a de novo automaton B′. In general, it is not possible for A′ and B′ to be equivalent. It is perfectly possible that A′ will fall into an adaptive crevice, while B′ will not. It is perfectly possible that A will, in general, fail to satisfy the threshold condition $n \geq 1$, whereas B and B′ will satisfy this condition. The material organization of a system is thus the critical factor that determines whether or not it is a suitable substrate for evolution or, more generally, for any learning process that depends on structural change.

Computers that can be decomposed into elementary machines are structurally programmable. Any systems to which the structure theory of automata applies are structurally programmable. The states of the elementary machines and their interconnectivity codes the program that generates their behavior according to a finite, well-defined set of rules. Such machines are unsuitable for evolution for the same reason that computer programs are unsuitable for evolution. They are overly delicate to structural change (unless they are fault tolerant, in which case they are overly insensitive to structural change for evolution to occur).

Biological organisms are quite different in this respect. Clearly, we can decompose biological systems into cells, such as neurons, into biochemical reactions that control the neuron, and into enzymes that control the biochemical reactions. But all these components combine aspects of continuity, bifurcation, and milieu dependence in a way that makes it impossible to define their potential interactions by a small set of well-defined rules. The folding of a protein enzyme to assume a three-dimensional shape and switching function is a continuous process that deforms gradually in response to mutations of a substantial fraction of the constituent amino acids. If it does not deform gradually enough, redundant amino acids can always be added that buffer the effect of mutation on features of shape critical for function. The biochemical reaction network controlled by the enzyme also combines gradual deformability, continuity, threshold behavior, and redundancy.

The firing of the neuron is an example of the threshold aspect. The conditions under which the neuron fires can change gradually as a consequence of changes in its internal biochemical dynamics or as a consequence of changes in the bath of hormones and other chemicals that surround it in the brain. As a consequence, collections of neurons also possess the evolution-facilitating features of continuity, bifurcation, milieu dependence, and function-buffering redundancy. We can think of the state-to-state behavior of each of these systems as described by a next state and output function, just as we can think of the state-to-state behavior of a structurally programmable finite automaton (e.g., a digital computer) as determined by a next state and output function. The key point is that when a genetic engineer wishes to ascertain how these functions change, he must either consult the laws of physics or perform an experiment. By contrast, the digital computer engineer precludes unexpected interactions and hides elements of continuity to the point where it is possible to choose desired next state and output function using well-defined rules listed in a programmer's manual of manageable size. This is the reason why biological materials are so much more malleable as substrates for evolution than are structurally programmable machines.

These considerations tell us that we cannot expect naked digital computers to evolve (supposing they could reproduce themselves). More to the point, we cannot expect computer programs to evolve at the level at which they are written, since they have all the features associated with well-defined rules of a programmer's manual at this level. According to Turing-Church, however, we can use the digital computer to simulate malleable organizations of the type that facilitate biological evolution. The structural changes (e.g., mutations) must occur at the level of the simulation. In other words, to make progress in the realm of evolutionary programming, it is necessary first to recognize the necessity of building a virtual machine with features of continuity, bifurcation, milieu dependence, and mutation buffering through redundancy. Virtual machines built on top of the base machine are ubiquitous in computer science. The time sharing systems built by operating systems designers are virtual machines. AI researchers often proceed by building a virtual machine with primitive operations suitable for a particular world, such as a blocks world. As we have seen, the problem for the AI researcher is to build a virtual machine to which he can communicate a suitable representation of the external world. The problem for the evolutionary programmer is to build a virtual machine to which he can communicate a suitable representation of the internal world of biological structure-function relations (see Conrad et al., Chapter 11, this volume).

If computing resources were infinite, we could look forward to eventually achieving open-ended evolution comparable to that exhibited by biological species. The reality is that resources are limited, and as a consequence, we must consider the computational costs of representing evolution-facilitating structure-function relations of a structurally nonprogrammable system. The componentry of biological organisms is organized for evolution; and through evolution, it can become specifically organized for the performance of particular information-processing tasks. The digital computer is universal. This means it can do anything, but it cannot do any particular thing as well as a special purpose system specifically tailored for the particular thing in question. It is not organized for evolution, and its components cannot be tailored for particular tasks through evolution. The situation is summed up by a tradeoff principle that ties together structural programmability, evolutionary adaptability, and computational costs (Conrad 1983, 1985). Roughly speaking, we can say that a structurally programmable system can achieve evolutionary adaptability only to the extent that computational resources are allocated to simulating structural nonprogrammability.

High-level AI also has the problem of paying for its representations in terms of computational resources. Evolutionary programming may have an advantage in this respect. Since the virtual system that is evolving is not programmable, sequentiality is unnecessary and not even desirable. Evolutionary programmming systems lend themselves to parallelism, and some of this effective recruitment of parallel resources can help to defray the costs of simulating nonprogrammability.

Evolutionary Architectures

So far, I have looked at evolution from the point of view of machines, starting with some presuppositions of computer science, proceeding to the evolution of computing, and trying to show how this evolution naturally leads to the concept of evolutionary computing and evolutionary programming. Now I would like to turn the situation around and look at machines from the point of view of evolution.

Imagine that we build a virtual evolutionary system, and that we evolve it to perform a particular task by means of selection. The situation is analogous to breeding plants or animals through artificial selection. Conceivably, we will eventually be able to develop systems that can continue to improve. We could harvest such systems from time to time. After some time—days, months, decades, even centuries— we might produce virtual machines that are quite useful for a particular task, such as a pattern recognition task. We might never understand how these systems work. Their dynamics might become altogether too complicated. But once evolved we would know all the parameters that control their dynamics and we could thus mass produce and distribute copies of the software as needed. We could never be sure that this software would behave exactly as we wish, just as we cannot be sure that plants, animals, and persons will behave as we wish. We also can't be sure a conventionally programmed computer will behave as we wish, and we can be pretty sure that error will creep into the program as it becomes larger. So, eventually, humans may come to accept programs with a self-organizing capability as not being intrinsically less trustworthy than the programs they design.

We can't expect our virtual evolutionary machines to have as much capability for self-organization as biological species. This is a consequence of the tradeoff principle. We have to reckon the costs of simulating nonprogrammability. Eventually, we might consider special architectures for evolution. These would probably include dynamical, analog-type hardware and possibly chemical hardware. After all, for the purposes of evolution, programmability is a nuisance. If we are going to work so hard to eliminate programmability at the virtual level, we might as well eliminate it in the base machine. So, we can imagine that evolutionary machines will come to incorporate some quite exotic architectural features. We wouldn't want these features in a universal digital computer. A universal digital computer is a formal system realized in matter, such as in silicon. As soon as we add in exotic dynamical features, we lose the power of programmability. So, we can imagine an eventual bifurcation of architectures. We will always want universal machines, and one family of architectures will continue to be associated with these. The second family of architectures will be associated with evolutionary machines. Universal machines are well suited for implementing high-level algorithms, while evolutionary machines will probably be well suited for adaptive pattern processing and other low-level adaptive processes. So, it is natural to imagine hybrid machines with a universal part and evolutionary parts. The evolutionary parts could be used for preprocessing ambiguous features of the environment, and the universal part for processing the disambiguated information.

All this is, of course, very speculative. Evolutionary programming does not yet have these kinds of dramatic achievements. My argument is that this lack of achievement is due to the presupposition of structural programmability that has led to the division of computer science thinking into language (software) and structural (hardware) branches. Progress in evolutionary programming has been inhibited by this presupposition. An evolutionary program consists of two parts. The first is the variation and selection search algorithm per se, and the second is the substrate on which this search algorithm acts. A good variation and selection search algorithm is important. A suitable substrate is essential.

An Evolutionary Look at the Fundamentals

Adding an open-ended capability of hardware evolution to machines raises some interesting questions about the fundamentals of computer science. We have assumed that we can use digital computers to model any process in nature. Since the universal digital computer is a physical realization of a formal process, its material embodiment is irrelevant. The computer model has the same relationship to the phenomenon modeled as does a mathematical description. It is not the same as the system modeled, but we assume that nothing essential is lost by virtue of the fact that we are dealing with codes rather than physical reality. This assumption, encapsulated in the Turing-Church thesis, has a long philosophical history.

We can clearly understand that choosing the representation has a bearing on computational efficiency. For example, an electron and a proton naturally attract each other. The digital computer that models this interaction must consist of a very large number of electrons and protons. It is clear that the choice of representation is all important here for the efficiency in terms of the number of particles necessary to realize the electron-proton interaction. The modeled system provides the most efficient representation. The digital computer is much less efficient, but it can be used to represent any system.

Possibly representation is important over and above efficiency. Wittgenstein (1953) argued that a symbol, such as a "pointing" finger on a roadsign, has no intrinsic interpretation. Interpretation depends on the larger system in which it is embedded. It is not clear that the problem of interpretation could be solved by embedding such a symbol in a larger system of symbols. The interpretation of the significance of the symbols would have to emerge from the interactions among the symbols. But, by definition, symbols do not admit of interactions. Computer programs are ordered collections of symbols, and one may ask how it is possible to interpret programs and their actions. Computer programmers do not worry about this problem, because it is they who are doing the interpreting. The semantical aspect of programming is parasitic on human intelligence. If Turing-Church is correct, it should be possible to produce a self-interpreting Turing machine (or, more generally, a self-interpreting process on a digital computer). If Wittgenstein is correct, this should be impossible.

Wittgenstein's argument does not apply to real evolutionary systems. Presumably, the phenomenon of interpretation could arise from the physical interactions among real-world entities such as atoms and molecules that are suppressed in purely formal systems used to represent these entities. Real evolutionary systems could have interpretive capability, whereas virtual ones could not. If this line of thought is correct, some further revision of presupposition is necessary. We would have to admit that evolutionary systems are more powerful than Turing machines, and that Turing-Church would fail with respect to the process of interpretation. Artificial intelligence on programmable machines would be limited in a fundamental way, over and above the limitations imposed by the availability of computational resources. Evolutionary machines would capture an aspect of cognition not captured in the formal computer paradigm.

Bibliography

Bremermann, H.J., 1962: Optimization through evolution and recombination, in Yovits, Jacobi, and Goldstein (eds.), *Self-Organizing Systems*, Spartan Books,

Burks, A.W., and A. Burks, 1981: The ENIAC—first general purpose electronic computer, *Annals of the History of Computation*, 3:310–389.

Conrad, M., 1972: Information processing in molecular systems, *Currents in Modern Biology* (now *BioSystems*) 5:1–14.

Conrad, M., 1981: Algorithmic specification as a technique for computing with informal biological models, *BioSystems* 13:303–320.

Conrad, M., 1983: *Adaptability*, Plenum Press, New York.

Conrad, M., 1985: On design principles for a molecular computer, *Communications of the ACM* 28:464–480.

Conrad, M., and A. Rosenthal, 1980: Limits on the computing power of biological systems, *Bulletin of Mathematical Biology* 43:59–67.

Conrad, M., and M. Strizich, 1985: Evolve II—a computer model of an evolving ecosystem, *BioSystems* 17:245–258.

Fogel, L., A. Owens, and M. Walsh, 1966: *Artificial Intelligence Through Simulated Evolution*, Wiley and Sons, New York.

Holland, J.H., 1975: *Adaptation in Natural and Artificial Systems*, University of Michigan Press, Ann Arbor.

Kampfner, R., and M. Conrad, 1983: Computational modeling of evolutionary processes in the brain, *Bulletin of Mathematical Biology* 45:931–968.

Kirby, K., and M. Conrad, 1986: Intraneuronal dynamics as a substrate for evolutionary learning, *Physica D.* 22:83–89.

McCulloch, W., and W. Pitts, 1943: A logical calculus of the ideas imminent in nervous activity, *Bulletin of Mathematical Biophysics* 5:115–133.

Minsky, M., 1961: Steps toward artificial intelligence, *Proceedings of the Institute of Radio Engineers* 49.

Rechenberg, I., 1973: *Evolutionsstrategie, Optimierung Technischer Systeme nach Prinzipien der Biologischen Evolution*, Friedrich Frommann-Verlag (Gunter Holzbook K.G.), Stuttgart-Bad Canstatt.

Rizki, M., and M. Conrad, 1985: Evolve III—a discrete events model of an evolutionary ecosystem, *BioSystems* 8:121–133.

Turing, A.M., 1936: On computable numbers with an application to the Entscheidungsproblem, *Proceedings of the London Mathematical Society*, series 2, 42:230–265.

von Neumann, J., 1956: Probabilistic logics and the synthesis of reliable organisms from unreliable components, in C.E. Shannon and J. McCarthy (eds.), *Automata Studies*, Princeton University Press, Princeton, N.J.

Wittgenstein, L., 1953: *Philosophical Investigations*, translated by G.E.M. Anscombe, Macmillan, London.

TABLE 10.1. Evolution of Programming*

A. Problems that computer scientists have either solved or
learned to manage

　　1) Machines and machine language
　　2) Fault tolerance (temporary reprieve through solid-state
　　　 technology and error detection-correction schemes)
　　3) High-level languages, compilers
　　4) Operating systems, timesharing
　　5) Resource requirements of some important algorithms
　　　 (classification into polynomial time, exponential time,
　　　 and NP-complete problems)
　　6) Increased computing resources (e.g., VLSI technology)
　　7) Structured programming
　　8) Storage and retrieval of large data bases
　　9) Distributed computing (messengers, protocols, guards)

B. Outstanding problems that are partially solved or
unsolved ("tough nuts")

　　1) Specification languages, flexible and adaptable code
　　2) Parallelism
　　　a) Efficient use of resources
　　　b) Attempts to contain the breakdown of programmability
　　3) Program correctness and fault tolerance of large-scale
　　　 computations
　　4) AI (e.g., pattern and object recognition, problem
　　　 solving, natural language processing, automatic
　　　 programming)
　　5) Learning and self-organization (presently, source of
　　　 intelligence is programmer)
　　6) Social problems associated with the delicacy of code
　　　 and the requirement of AI for unambiguous environments

*The order very roughly reflects the successional interplay
of problems, solutions, and new problems created by these
solutions, including problems and solutions involving
hardware developments and application needs. In fact, these
developments had a highly concurrent aspect.

TABLE 10.2 Evolutionary Programming

1) Specify goals and evaluation criteria

2) Machine learns to achieve goals by variation and selection (or trial and error) algorithm

3) Error is useful (some program incorrectness and faults are necessary)

4) Sequentiality not necessary or even desirable; machine can learn how to use parallel resources efficiently

5) Learning makes it possible to deal with variable and ambiguous environments; evolutionary learning is a fundamental form of adaptability

6) Source of intelligence is inside system (programming is truly automatic)

Michael Conrad, Roberto R. Kampfner,
Kevin G. Kirby

11. Neuronal Dynamics and Evolutionary Learning

Introduction

Evolution by variation and selection is nature's foundational method of problem solving. But it is not a universal method. The organization on which variation and selection acts must be suitable. Some organizations, such as raw computer code, are too delicate. What makes proteins, cells, and organisms suitable is the malleable relationship between their structure and their computational function.

We have developed two simulation systems for studying evolutionary learning that explicitly address the structure-function issue. Both systems are based on the idea that patterns of neural signals impinging on the neuron are transduced to activity patterns, such as chemical patterns, that are detected by enzymes. Each such neuron has two levels of dynamics. The upper level is associated with the transduction of the input signal pattern to a chemical pattern, while the lower level is associated with the shape-based molecular pattern recognition at the level of the enzyme (Conrad 1985). In these models, the brain is viewed as a network of neurons, many of which are themselves endowed with powerful pattern-recognition capabilities.

The first of these two systems, which we will call Learn I, utilized discretized models of these powerful pattern-processing neurons (Kampfner and Conrad 1983). The second model, which we call Learn II, models the chemical processes that mediate the conversion of pre-synaptic input patterns to chemical patterns within the neuron in detail (Kirby and Conrad 1984). This chapter will focus our attention on Learn II. We will briefly describe the biological motivation for the model, and outline the structure and operation of the simulation system. Our main aim, however, is to illustrate how an information-processing system structured for effective evolution can be represented in a computer program. We will describe a small but relevant cross-section of the experimental results obtained with the system, and briefly characterize its computational capabilities.

The Conceptual Model

The evolutionary selection circuits model (Conrad 1974, 1976) proposes the successive modification of enzymatic neurons through a process of variation and selection as a means of developing new information-processing functions in the brain. Important biochemical correlates of enzymatic neurons have already been identified (Greengard 1978; Liberman et al. 1982). This fact, and results from simulations with Learn I, strongly suggest the feasibility of evolutionary mechanisms of the type postulated in this model.

We will describe the system in a top-down fashion. Learning proceeds by modification of the information-processing functions of enzymatic neurons. These are neurons whose firing behavior is controlled by enzymes, to be called *excitases*. As we shall see, enzymatic neurons have powerful pattern-processing capabilities as compared to simple threshold neurons. The main hypothesis is that the gradual transformation of these capabilities can be accomplished through the successive application of the following three-step evolutionary cycle to a population of competing systems:

1. The performance of the existing structures is evaluated on the basis of each system's interactions with the environment. Such structures are defined by the excitase composition of the systems.
2. Traits are propagated from the systems showing the highest performance to other systems in the population. By a system's traits we mean the excitase composition of each of its enzymatic neurons.
3. The excitase composition of the systems is subject to random variation. In the formal model, this amounts to adding to (or deleting from) each enzymatic neuron with a specified probability, randomly selected excitase types. In natural systems, or in actual implementations of the model, new excitase types would normally result from random mutations affecting the primary structure of protein enzymes. Therefore, their detailed characteristics and functional properties cannot be predetermined.

Evolutionary learning calls for the gradual adaptation of information-processing functions until a specified behavior is acquired. As already mentioned, enzymatic neurons incorporate the principle of double dynamics (Conrad 1985), which allows for the transduction of electro-chemical signals into a chemical pattern that excitase enzymes can recognize. In discrete neurons, both levels of dynamics are expressed by a simple rule that defines the correspondence between excitase types and the input patterns to which each one responds. In continuous (reaction-diffusion) neurons, receptor molecules activated by transmitters trigger the production of cAMP. This produces a sudden "jump" in cAMP concentration at the membrane sites where the dendritic inputs are located. The initial pattern of cAMP concentration diffuses on the neuron's membrane, and is subsequently interpreted by excitases, typically protein kinases, which trigger the neuron's response. A continuous enzymatic neuron fires whenever an above-threshold concentration of cAMP is reached in a site on the neuron's membrane where a kinase is located. Excitases are identified with kinases, and thus constitute the pattern-recognition primitives. Other forms of dynamics have been defined for enzymatic neurons (Kirby and Conrad 1986). However, this chapter will concentrate on results obtained from simulations with continuous (reaction-diffusion) neurons.

We model an enzymatic neuron as a grid, or tessellation structure, with n compartments $k = 1, 2, \ldots, n$; m input lines, x_i, $i = 1, 2, \ldots, m$, corresponding to presynaptic inputs; and one output line, y, that corresponds to the axon in a biological neuron (see Figure 11.1).

Formally, the firing rule and the underlying dynamics are expressed as follows: The excitase configuration of the neurons can be represented as a constant binary vector $z = [z_1, z_2, \ldots, z_n]$, where the subscripts refer to compartments on the cell membrane. $z_k = 1$ means that an excitase enzyme is located at compartment k. Excitase enzymes are activated by an above-threshold concentration of cyclic nucleotides in the compartment in which they are located. Thus, at any time t, the output of the neuron y(t) is determined by its excitase configuration z in conjunction with an internal state vector $u(t) = [u_1(t), u_2(t), \ldots, u_n(t)] \in R^n$, which describes the level of excitation (i.e., the cyclic nucleotide concentration) at each compartment. If the global threshold concentration is Θ, the firing rule for an enzymatic neuron is

$$y(t) = H[\max z_k u_k(t) - \Theta] \tag{1}$$

where H is the unit step function. A continuous neuron is in fact a reaction-diffusion network of cyclic nucleotides (Kirby and Conrad 1984). Here, the internal state vector is determined by the reaction-diffusion equation

$$du_k(t)/dt = d_{(ik)}[u_i(t) - u_k(t)] + R_k[u_k(t)], \tag{2}$$

and by discontinuous dynamics of inputs and firing

$$u_k(t + dt) = Ix_k(t) + u_k(t)[1 - ay(t)], \quad 0 < a < 1. \tag{3}$$

An important characteristic of this dynamics is that similar input patterns are likely to produce similar internal states, and that there is also a feedback loop from the firing (y) to the internal state (u). The consequence is that each excitase is associated with a family of related input patterns instead of a single one.

The enzymatic neuron model has at least two layers of information processing that underlie the classical electrical layer. The first is the reaction-diffusion dynamics of cyclic nucleotides. This mediates the transduction of incoming signal patterns into cAMP concentration patterns. The excitases which read out these concentration patterns constitute a second layer of information processing at the molecular level. Standard connectionist models rely mainly on the electrochemical layer of nerve impulse activity for information processing, and it should be clear that the enzymatic neuron model is different and potentially more powerful in this respect.

The Learn II Simulation System

The structure of the Learn II simulation system is depicted schematically in Figure 11.2. Modules VARIATION, SELECTION, and PROPAGATION incorporate the basic functions of the selection circuits model. Modules TASK, INTERFACE, and NEURON are accessed by the SELECTION module. The module TASK monitors the interactions of the simulated robots with the environment as the task is performed in each evolutionary cycle. Module MAIN integrates the functions of these modules and monitors the simulation of the evolutionary learning process as a whole. The Learn II simulation system allows for the evolutionary learning of a variety of adaptive pattern-recognition tasks, using enzymatic neurons with different dynamics and varying the parameters that control the variation and selection process. The Learn II simulation system was implemented on Wayne State University's Amdahl 470/V8 and on the VAX 11/780 machine in our department. The experiments about gradual acquisition of vector fields were conducted on the VAX. The rest of the robot navigation experiments were conducted on the Amdahl.

A Robot Navigation Task

We have analyzed the adaptive and computational capabilities of systems based on continuous enzymatic neurons using a robot navigation task. In this type of task, a robot moving on a plane must reach a fixed target in the presence of randomly directed mechanical forces. To perform this task, the robots start from an initial position, and must reach the target within a specified number of steps. In the simulations, we identified the target with the origin of the coordinate system, and specified the initial position by its cartesian coordinates. Figure 11.3 illustrates a possible trajectory of a robot that performs the task successfully. Typically, a robot starting at initial position P_1, with coordinates (7,7), say, is supposed to reach the target within five steps.

Each robot is controlled by a network of four continuous enzymatic neurons. As illustrated in Figure 11.4, at each step during the execution of the task, the performing robot receives information about (1) the relative position with respect to the target, (2) the distance from the target, (3) the direction of the mechanical forces acting on it, and (4) the magnitude of such forces. Table 11.1 describes the coding scheme used in the simulations for these inputs. The distance of the performing robot to the target at any step during the execution of the task is computed as the euclidean distance from the robot's position to the origin of the coordinate system.

At each step, the robots also face randomly directed mechanical forces of varying magnitudes. Random forces detected by a robot at step t, for example, are assumed to take effect at step t + 1. Table 11.2 gives the interpretation of firing frequencies in terms of displacements caused by the random forces. The robot's action at a given time step equals the composition of individual actions, or "moves," in any of four possible directions, produced by each of the four effectors controlled by the robot's control systems. These actions are the responses of the robot control systems to input patterns encoding environmental stimuli. The position of the robot at the end of each step is, therefore, the result of the composition of two displacements: one caused by its own action, and the other by the random forces. Outputs from the networks encode actions to be taken by effectors in the robots at each time step. Table 11.3 describes the coding scheme for network outputs used in the experiments. The interpretation of the corresponding frequencies is illustrated in Table 11.4.

Experiments on Robot Navigation

Robot navigation in a quiet environment. In a quiet environment, the robots are not subjected to the action of randomly directed mechanical forces. Therefore, in this case, the information concerning the direction and magnitude of the random forces is not required by the robots. There are eleven possible values per input line, so this reduces the number of possible environmental inputs from 11^8 (on eight input lines), required when the robots face a noisy environment, to 11^4 (on four input lines). The task of robot navigation in a quiet environment, which is the one described in Figure 11.3, was learned in an average of 5 evolutionary cycles. These results clearly confirm the adequacy of systems based on continuous neurons for the development of specific responses to patterns of environmental stimuli, so as to achieve a pre-specified goal.

Robot navigation in the presence of random forces. When the robots face randomly directed mechanical forces at each step of the execution of the task, the uncertainty of the environment increases considerably. The state of the environment is, in this case, determined jointly by the position of the robot on the grid and the direction and magnitude of the forces. But the latter varies randomly. Therefore, the state of the environment at the next step is not completely determined by the action of the robots. However, our experiments showed that evolutionary learning is remarkably efficient even under these conditions. It took only 10 evolutionary cycles, on the average, for a robot based on continuous neurons to learn the task described in Figure 11.5. This means that evolutionary systems of the kind discussed here are also apt for learning tasks in which the systems face a relatively uncertain environment. It is interesting to note that robots facing a noisy environment sometimes took advantage of mechanical forces acting on them, thus exhibiting a gliding behavior reminiscent of some simple biological organisms. The efficiency of evolutionary learning further confirms the crucial role that evolutionary strategies and mechanisms play in the adaptation of nonprogrammable systems. We will discuss important aspects of this role later. In particular, we discuss the issue of constraints and dynamics and the role of generalization in evolutionary learning.

Learning to reach the target from each of the four quadrants. Training of the robots was conducted successively in each of the four quadrants, with a pre-specified initial position. Random forces were also present in these experiments. The major challenge here is for the robots to develop excitase configurations that would allow them to perform well in all the quadrants. Figure 11.6 shows the paths followed by a robot in each of the four quadrants at an advanced stage of the learning process. This task shows clearly some of the constraints imposed by the dynamics on the possible behaviors of the systems.

Capabilities of the System

In the following sections, we discuss important features of evolutionary learning that became apparent in our simulation experiments with Learn II.

Gradual acquisition of vector fields. The excitase configurations of the four neurons in the robot correspond to a vector field on the plane. The vector at position (x,y) has length

$$(dx^2 + dy^2)^{1/2},$$

where dx and dy are the number of steps the robot will take in the x and y directions, respectively. Even though we must use a discretized plane in practice, the coding scheme described previously is valid for any real x and y, so we have a general field

$$v: R^2 \longrightarrow R^2.$$

As the excitase configurations are mutated, the field changes. We have found that we can mathematically characterize the repertoire of vector fields possible for a robot using enzymatic neurons as a substrate. The analysis is presented elsewhere (Kirby and Conrad 1986), and here we only show how learning can be viewed as the gradual "acquisition" of a certain vector field.

Figure 11.7 shows how the vector fields evolve in the course of learning. After one cycle, we have the field shown in Figure 11.7a. In the picture, all vectors are normalized to unit length for readability. The x's correspond to vectors with length zero. The diagrams at the bottom are abbreviated notations for the excitase configurations of the four neurons, causing (from right to left) motion in the N,S,W,E directions. At this level of abstraction, excitase enzymes are the same if they produce the same output behavior. In particular, we place a mark in the large circle corresponding to a certain direction if there is an enzyme near enough to the synapse for position sensing in that direction to go above threshold. The small circles represent the presence or absence of excitases in membrane regions where the effects of two position inputs overlap. For example, in Figure 11.7a, there is one excitase in the North motor neuron in a region of the membrane sensitive to simultaneous inputs on the south and west input lines; in the West motor neuron, we have the same situation; the other two neurons are empty of excitases.

Figure 11.7a was obtained after the end of the first learning cycle. It shows that the robot will find the target only if it is placed in the third quadrant on the line x = y. After six learning cycles, the vector field has changed to that of Figure 11.7b. In this case, robots in all four quadrants get closer to the target, and they can find it exactly if placed anywhere in the first quadrant. Finally, after twelve learning cycles, the task has been learned. The resulting vector field is shown in Figure 11.7c. No matter where the robot is placed in the plane, it will move toward the origin.

The important point is that the target is found by training the robot from only four test positions, and the evaluation is purely local. Yet, the end result is globally the best. This shows that the dynamics of the system is very well matched to the task.

Constraints and Continuity

The lower level of dynamics in continuous neurons is mediated by the patterns of diffusion of cAMP concentration. From equations (1) to (3), it can be seen that many distinct input patterns may cause an above-threshold concentration in a given compartment of the neuron's membrane. Of course, this feature makes the patterns similar—that is, belong to the same class. Both time and space features of the patterns may be involved in this similarity. We refer to the ability of enzymatic neuron-based systems to produce the same response to a set of input patterns as their generalization capability. In discrete neurons, this capability does not originate in the properties of individual excitases. Rather, it is determined by the excitase configuration of the neurons of the system and it is, therefore, developed in a step-by-step fashion through the evolutionary process. In continuous neurons, on the other hand, the diffusion dynamics allows each excitase enzyme to respond to a family of input patterns instead of a single one. This implies that families of recognized inputs can change in a single evolutionary cycle, as opposed to the single change in patterns recognized by discrete neurons.

We shall refer to the notions of performability and learnability of tasks by enzymatic neuron-based systems in order to discuss the issue of constraints and continuity. A task is said to be *performable* by an enzymatic neuron-based system if there exists an excitase configuration that produces an adequate response to each input pattern that may arise during its execution. Clearly, not all the possible input patterns can be distinguished by an enzymatic neuron endowed with a specific type of dynamics. Thus, this type of constraint relates to whether a given task can be performed by a system having a specific continuous dynamics.

A different type of constraint affects the ability of systems of this type to learn a given task. We say that a task can be *learned* by an enzymatic neuron-based system using evolutionary mechanisms if there exists a sequence of excitase configurations, each element of which is obtained from its predecessor by mutation, which terminates in a configuration for the successful performance of the task. As a consequence of the selectional nature of the algorithm, performance will increase monotonically in this sequence. We say that a task is learned *efficiently* when evolutionary learning does better than random search. Clearly, a task may be performable by an enzymatic neuron-based system, and yet it might not be possible for this system to learn such task efficiently. In a sense, we can say that a system's generalization capabilities amount to its ability to "anticipate" certain features of the environment. But it is the evolutionary process that ensures that the acquired generalization capabilities serve a useful purpose. Of course, this only happens when, for a specific dynamics, the task is both realizable and amenable to evolutionary learning.

Conclusions

Our experience with the robot navigation task (and with related pattern-recognition tasks not described here) suggests that evolutionary programming is a viable approach. But as stated at the very beginning of our discussion, variation and selection is not a universally effective optimization technique. It works only if the organization on which it acts is reasonably likely to undergo useful variations. Enzymatic neural nets are structured to fulfill this condition.

This structuring entails computational costs. It is evident that we have had to simulate biology-like structure-function relations in order to make evolution work effectively. To do this, we built a virtual machine on top of the base machine and allowed variation and selection to operate at the virtual level. Setting up a virtual machine is quite common in AI programming. And the usual situation is that such virtual machines are inefficient due to the fact that they are unsuited to the hardware on which they run. The distinctive feature of our virtual machine is that it is highly dynamical. It would thus lend itself to parallel implementations in which each neuron is simulated with a separate processor, with each processor running for approximately the same amount of time. The system also lends itself to parallel implementation due to the fact that conventional programmability is not important. Finally, once a useful "brain" is evolved, it could be hardwired for repetitive use; at this point, the overhead of the variation-selection algorithm could be dispensed with unless continued evolutionary adaptability is required.

Our model has one major point of similarity to connectionist models of cognition and one major point of contrast. The point of similarity is that structure is important for the function of a cognitive system. The point of contrast is that computation and cognition are not due in the main to connectivity among simple components. Structures and dynamical processes within cells play the more important role. Connectivity in the selection circuits model would serve mainly to orchestrate suitable combinations of differently specialized enzymatic neurons and to mediate memory storage and retrieval processes (Conrad 1976).

The computer study described here shows that dynamically based information processing has sufficient plasticity to satisfy the central requirement of a brain model—that it have enough plasticity to actually develop in the course of evolution. Important forms of ontogenetic learning probably depend on this plasticity as well. The computational costs of simulating plasticity would impose severe limits on the cognitive capabilities of conventional architectures and connectionist models in which neurons are represented as simple switches. But to the extent that we can capture some of this plasticity with well-designed simulations and suitably designed hardware, it should be possible to bring some new important problems into the purview of artificial computing systems.

Bibliography

Conrad, M., 1974: Evolutionary learning circuits, *Journal of Theoretical Biology* 46:167–188.

Conrad, M., 1976: Complementary molecular models of learning and memory, *Biosystems* 8:119–138.

Conrad, M., 1985: On design principles for a molecular computer, *Communications of the ACM* 25:5.

Greengard, P., 1978: *Cyclic Nucleotides, Phosphorylated Proteins, and Neuronal Function,* Raven Press, New York.

Kampfner, R., and M. Conrad, 1983: Computational modeling of evolutionary learning processes in the brain, *Bulletin of Mathematical Biology* 45:931–968.

Kirby, K.G., and M. Conrad, 1984: The enzymatic neuron as a reaction-diffusion network of cyclic nucleotides, *Bulletin of Mathematical Biology* 46:765–782.

Kirby, K.G., and M. Conrad, 1986: Intraneuronal dynamics as a substrate for evolutionary learning, *Physica D.* 22:205–215.

Liberman, E.A., S.V. Minina, N.E. Shlovsky-Kordy, and M. Conrad, 1982: Microinjection of cyclic nucleotides provides evidence for a diffusional mechanism of neuronal control, *Biosystems* 15: 2, 127–132.

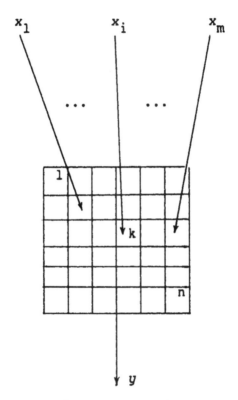

FIGURE 11.1 An enzymatic neuron. An input pattern produces
a pattern of excitase concentration on the compartments of
the neuron's membrane. This pattern of concentration changes
over time due to reaction-diffusion processes. The neuron
fires if an above-threshold concentration exists in a compartment where an excitase is located.

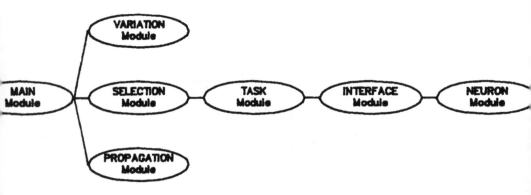

The structure of the Learn II simulation system.   The modules are indeper
allowing each to be replaced with alternate versions.

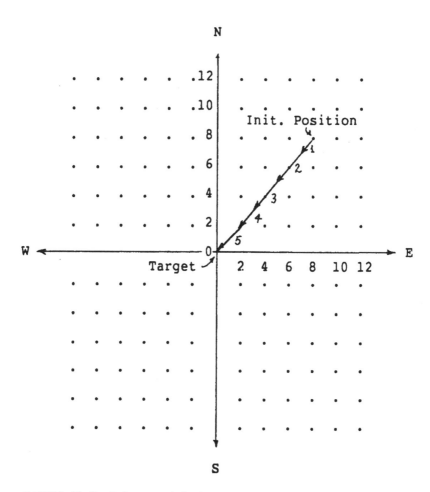

FIGURE 11.3 Robot navigation in a quiet environment. The figure shows the route followed by a trained robot that reaches the target in the five steps allowed.

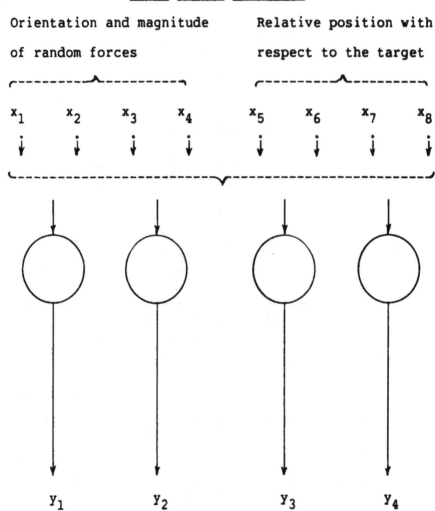

Input Vector Positions

Orientation and magnitude
of random forces

Relative position with
respect to the target

x_1 $\quad$ x_2 $\quad$ x_3 $\quad$ x_4 $\qquad$ x_5 $\quad$ x_6 $\quad$ x_7 $\quad$ x_8

y_1 $\qquad$ y_2 $\qquad$ y_3 $\qquad$ y_4

$y_1, y_2, y_3, y_4 \in [0,n]$ $\qquad\qquad$ $x_1, x_2, \ldots, x_8 \in [0,m]$

FIGURE 11.4 A robot's control system. It consists of a
network of four enzymatic neurons. Each neuron's output
(y_i) controls a distinct effector. All neurons, however,
receive identical inputs x_j, $j=1,2,\ldots,8$.

TABLE 11.1 Input codes used in robot navigation task

	Input vector position		Firing frequency	
Number	Orientation of random forces	Relative position of robots	Magnitude of forces	Distance from target along axis
1	North		n	
2	South		n	
3	West		n	
4	East		n	
5		North		m
6		South		m
7		West		m
8		East		m

n ϵ [0,6], m ϵ [0,10].

TABLE 11.2   Action of forces

Magnitude of forces	displacement
$n > 3$	2
$1 < n \leq 3$	1
$0 < n \leq 1$	0

TABLE 11.3   Output codes in robot navigation task

Output vector position	Direction of action	Firing frequency
1	North	k
2	South	k
3	West	k
4	East	k

TABLE 11.4   Robot's actions

Frequency	Displacement
$k > 1$	2
$0 < k \leq 1$	1
$k = 0$	0

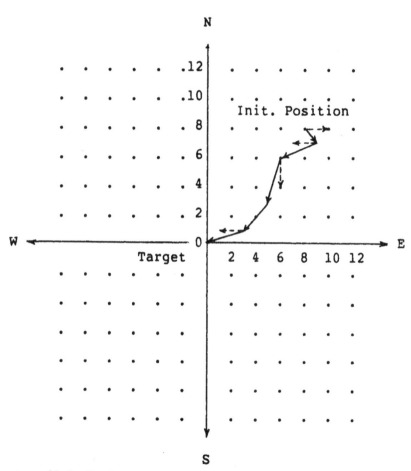

FIGURE 11.5 Navigation task with robots facing random forces.
The figure shows the route followed by a trained robot. The
direction and magnitude of the forces at each step is also
shown (broken arrows).

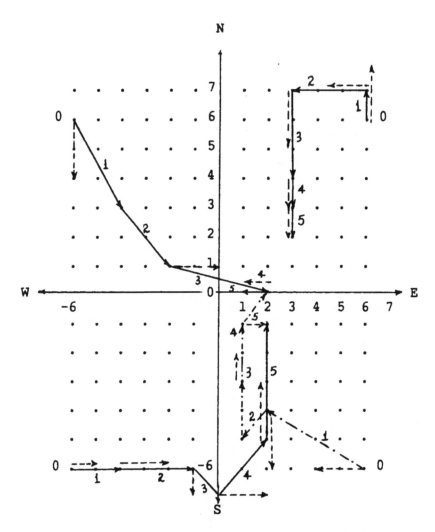

FIGURE 11.6 Routes followed by a robot in a task that consists of reaching the target from any of four initial positions, one in each quadrant, indicated by a 0. Broken arrows indicate direction and magnitude of random forces. The robot shows an advanced stage of learning.

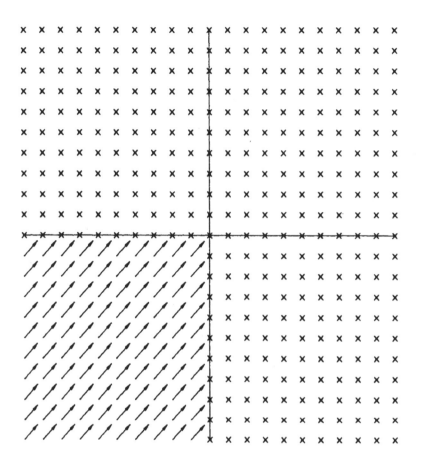

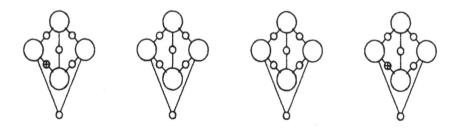

FIGURE 11.7a Gradual acquisition of vector fields by the motor control system: an initial state

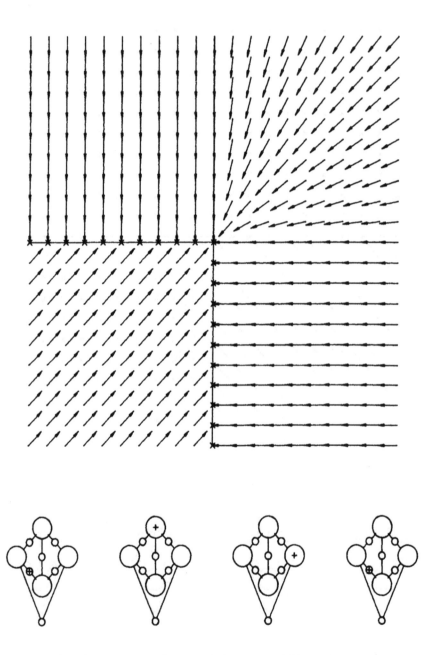

FIGURE 11.7b Gradual acquisition of vector fields by the motor control system: partial learning

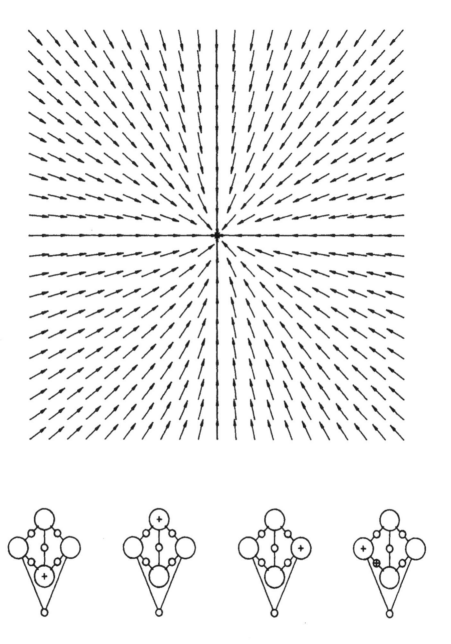

FIGURE 11.7c Gradual acquisition of vector fields by the
motor control system: total learning

12. Generalization in Evolutionary Learning with Enzymatic Neuron-based Systems

Introduction

Some experiments we conducted involving computational modeling and simulation of evolutionary learning processes with enzymatic neurons (Kampfner and Conrad 1983a) suggested the feasibility of evolutionary mechanisms for adaptive pattern recognition and adaptive motor control. An evolutionary learning algorithm was developed in connection with these simulation experiments, and its main parameters identified and analyzed. These parameters control the potential variability and the rates of variation to which the learning systems are subjected. These early experiments also showed that there is a definite relationship between the values of these parameters, and that settings of these parameter values which are associated with an increased efficiency of learning reflect conditions enhancing the gradualism necessary for evolutionary adaptation. The role of individual neurons as powerful information processors in neuronal networks has also been characterized and studied. In this respect, enzymatic neurons are viewed as pattern recognition elements organized according to the double dynamics principle which, essentially, allows for lower level dynamic processes to mediate the transduction of external signals received by the neurons into the corresponding neuronal response (Conrad 1985; Kirby and Conrad 1984, 1985). A hierarchy of neuron models incorporating the double dynamics principle has been proposed recently, and properties of two of these models have been studied (Kirby and Conrad 1985).

This chapter will discuss the role of generalization in adaptive pattern recognition and adaptive motor control processes. We refer to two types of neurons in the mentioned hierarchy, linear and cyclic nucleotide neurons, and relate their properties to the generalization concept. For the purpose of the present discussion, we define generalization as the potential ability of systems to abstract general features of the environments to which they have been exposed through learning. The learning systems are said to generalize on the basis of their current "knowledge" the extent to which they are able to handle new environmental situations having features in common with already known environments. Thus, in our context, generalization means essentially the ability of systems to correlate their responses with general features abstracted from the patterns included in a training set. The results reported here refer to tasks consisting essentially of pattern recognition/ pattern generation processes. For certain tasks of this type (e.g., tasks involving adaptive motor control), it is also possible to characterize generalization as a rule-learning process. We will use this characterization in the sequel. We do not attempt, however, to give the generalization capability any interpretation in terms of higher level cognitive processes. Our interpretation is rather in terms of some kind of primitive generalization capabilities. However, as will be explained, this kind of generalization appears to be extremely useful as a means of enhancing the evolutionary learning capabilities of enzymatic neuron-based systems.

This chapter is organized as follows. The evolutionary selection circuits model of evolutionary learning processes in the brain (Conrad 1974, 1976) and the essential features of discrete (a special case of linear neurons) and continuous (cyclic nucleotide) neurons are discussed in the second section. There, we also give a brief description of the adaptive pattern recognition and adaptive motor control tasks involved in the simulations (see also Chapter 11, this volume, for a detailed description). In the third section, the learning algorithm is described, and the learning process is characterized as a search in the space of functions performed by networks of enzymatic neurons. We also discuss the structure of the search space and the main factors affecting the search process. Evolutionary learning with discrete neurons, the concept of evolutionary generalization, and the role of gradualism in evolutionary learning are discussed in the fourth section. The impact of continuous information processing on evolutionary learning is then discussed in the fifth section. In this respect, the intrinsic generalization capability of continuous neurons and its relationship to concomitant constraints on the behavior of the systems are the main focus. The final section provides a summary of the role of generalization in evolutionary learning processes, and suggests some issues for further study and research.

Evolutionary Learning

A Conceptual Model of Evolutionary Learning

The evolutionary selection circuits model (Conrad 1974, 1976) describes learning as the result of the gradual modification of enzymatic neurons through a process of variation and selection. Enzymatic neurons are formal neurons whose behavior is mediated by excitase enzymes. The underlying principle of the behavior of enzymatic neurons is that excitase enzymes, activated by specific input patterns, cause the neuron to fire. In the evolutionary selection circuits model, learning involves the gradual modification of the excitase configuration in a population of enzymatic neurons until a desired behavior is acquired by at least one of the neurons in the population. The learning mechanism, which can be easily generalized to networks of these neurons, consists of the following steps:

1. evaluation of the performance of enzymatic neuron-based systems on the basis of their interactions with the environment,
2. selection of the best-performing systems on the basis of this evaluation, and propagation of traits (i.e., excitase enzymes) from these systems to others in the population, and
3. application of random variation to the systems not showing the best performance.

This process is repeated until a system showing the desired performance is found.

An important feature of this model is that the effectiveness of the learning process does not depend primarily on the efficiency of the search process, but on the amenability of the learning systems to gradual evolution. This chapter explores the relationship between the generalization capabilities of evolutionary systems, in the sense defined earlier, and their amenability to effective evolution.

Discrete and Continuous Neurons

The dynamics of enzymatic neurons operates at two main levels. An upper level describes their behavior in terms of a firing rule that relates the excitase configuration of the neuron to its internal state. The excitase configuration of the neurons can be represented as a constant binary vector $z = [z_1, z_2, \ldots, z_n]$, where the subscripts refer to compartments on the cell membrane. $z_k = 1$ means that an excitase enzyme is located at compartment k. Excitase enzymes are activated by an above-threshold concentration of cyclic nucleotides in a compartment in which they are located. Thus, at any time t, the output of the neuron y(t) is determined by its excitase configuration z in conjunction with an internal state vector $u(t) = [u_1(t), u_2(t), \ldots, u_n(t)]$ in R^n, which describes the level of excitation (i.e., the cyclic nucleotide concentration) at each compartment. If the global threshold concentration is Θ, the firing rule for an enzymatic neuron is

$$in \ y(t) = H[max \ z_k u_k(t) - \Theta] \qquad (1)$$

where H is the unit step function.

Both discrete and continuous neurons share the same upper level dynamics. A second level dynamics describes the evolution of the internal state of enzymatic neurons. It is at this level where the difference between continuous and discrete neurons can be explained. In the linear neuron (Kirby and Conrad 1985), weighting coefficients $w_{(ij)}$ specify the influence of input j on compartment i. Effects from different synapses add linearly, and are assumed to take effect at the same time in all compartments. Thus, in the linear neuron, the state vector at time t is defined by

$$u(t) = W x(t), \qquad (2)$$

where W is the weight matrix and $u(t)$ is a binary input vector that specifies the inputs that are active at each synapse. We refer here, as discrete neurons, to a special case of linear neurons where each excitase enzyme is activated by a unique input pattern (i.e., a unique input vector configuration). A continuous neuron, on the other hand, is in fact a reaction-diffusion network of cyclic nucleotides (Kirby and Conrad 1984). Here, the internal state vector is determined by the reaction-diffusion equation

$$du_k(t)/dt = d_{(ik)} [u_i(t) - u_k(t)] + R_k [u_k(t)] \qquad (3)$$

and by the discontinuous dynamics of inputs and firing

$$u_k(t + dt) = I x_k(t) + u_k(t)[1 - a y(t)], \ 0 < a < 1. \qquad (4)$$

An important characteristic of this dynamics is that similar input patterns are highly likely to produce similar internal states, and that there is also a feedback loop from the firing (y) to the internal state (u). The consequence for the firing rule of equation (1) is that, especially when firing behavior over time is considered, each excitase is associated with a family of related input patterns instead of a single one. According to this, we say that continuous neurons have the intrinsic generalization property, while discrete neurons do not. We will discuss the effect of this property on the process of evolutionary learning in the following sections.

Adaptive Pattern Recognition and Adaptive Motor Control

Enzymatic neurons are essentially pattern recognizers. Discrete neurons recognize the class of input patterns to which they respond by firing. A network of m discrete neurons can classify a set of input patterns into 2^m classes, the number of possible combinations of neurons in the network. Continuous neurons, on the other hand, naturally classify spatio-temporal patterns into classes defined by the frequency of their responses. Consequently, a network of m continuous neurons can classify a set of input patterns into a number of classes proportional to 2^m. In general, the number of pattern classes is greater for continuous neurons than it is for discrete neurons.

We will briefly describe the robot navigation task, a pattern recognition/pattern generation task in which a network of enzymatic neurons performs the motor control functions of the simulated robots. The network configuration is depicted in Figure 11.4 (see Chapter 11, this volume). In this task, the robots must learn to reach a fixed target within a pre-specified number of time steps. The robots start at an initial position on a grid. At each time step, they are subjected to the action of randomly oriented mechanical forces that displace them. The robots can take an action consisting of a move in any, or a combination, of four possible directions. It is assumed that the networks receive inputs encoding the direction of randomly oriented mechanical forces acting on the robots, and the relative position of the robots with respect to the target. For continuous neurons, however, the networks also detect the magnitude of the mechanical forces and the robot's distance to the target, both encoded as firing frequencies of the corresponding input lines (see Figure 11.4, Chapter 11, this volume). The outputs of the networks are interpreted as encoding instructions for effector actions to be taken in response to the inputs received at each time step. These actions correspond to moves in each of the four possible directions parallel to the coordinate axes (i.e., North, South, East, and West directions, with the usual convention). If movement is in more than one direction at any one time, the resulting action is the composition of the individual moves. Figure 12.1 shows a sequence of moves taken by a simulated robot with continuous neurons. In this sequence, the robot reached the target in the five steps allowed.

The Evolutionary Learning Process

The Learning Algorithm

The learning algorithm used in the simulations of adaptive pattern recognition and adaptive motor control (Kampfner and Conrad 1983a) incorporates the evaluation, variation, and selection processes of the selection circuits model. The basic functions of the learning algorithm are described schematically in Figure 12.2. The algorithm allows for a step-by-step variation of the structures represented—that is, the networks of enzymatic neurons. This is important for evolutionary learning. The performance of the task includes the simulation of the interactions of the systems with the environment and the evaluation of the degree of success achieved by the robots in their attempts to reach the target. The coding schemes used to encode environmental states as inputs to the networks (i.e., the direction and magnitude of the random forces, together with the relative position of the robots with respect to the target and the distance to the target from their current position) is given in Tables 11.1 and 11.2 (see Chapter 11, this volume). The interpretation of network outputs in terms of robot actions is illustrated in Tables 11.3 and 11.4 (see Chapter 11, this volume). The performance measure used was the euclidean distance between the final position of the robots after the allowed number of moves and the position of the target on the grid. The type of dynamics used is transparent to the learning algorithm per se. The dynamics module used in the simulations of continuous neurons was developed by Kirby (Kirby and Conrad 1984).

The Search Space

Evolutionary learning with enzymatic neuron networks is a search in the space of available excitase configurations. The size of this space depends on the amount of resources required by enzymatic neuron-based systems to perform a given task—that is, the number of excitase types required to configure a structure capable of performing the task.

Pattern recognition tasks with discrete neurons. Taken individually, discrete neurons classify patterns of binary digits into two classes. One class corresponds to those patterns in response to which the neuron fires. The other class corresponds to those patterns that do not cause the neuron to fire. As mentioned in the previous section on evolutionary learning, discrete neurons are a special case of linear neurons in which each excitase enables them to respond to a unique input pattern.

Therefore, to perform correctly a specific pattern recognition task, a discrete neuron must contain an excitase configuration with exactly those excitases corresponding to the patterns to be responded to by the neuron. There are 2^n distinct input patterns of n binary digits. Consequently, 2^n excitase configurations must be available for enzymatic neurons to learn to classify n-digit patterns. There are 2^{2^n} distinct dichotomies of a set of 2^n distinct patterns. With discrete neurons, each such dichotomy (that is, each distinct pattern recognition task) requires a specific excitase configuration. Thus, learning a pattern recognition task with (single-neuron) networks of discrete enzymatic neurons amounts to a search on a space of 2^{2^n} distinct excitase configurations.

Random search as a base line. Random search on the space of available excitase configurations is a useful base line for the efficiency of the learning process. Specifically, to find a structure that dichotomizes an n-binary digit pattern, choosing structures according to a uniform probability distribution, the expected number of trials is $2^{2^n}/n$. Thus, with random search, the number of trials needed to learn a task is an exponential function of problem size (in this case, 2^n, the number of patterns classified).

Factors Affecting the Search Process

Evolutionary learning is a search on a space of structures that define systems capable of performing specific tasks. We are concerned with structures representing single-layer networks of enzymatic neurons that perform the adaptive pattern recognition/pattern generation tasks defined previously. Implicit in the enzymatic neuron concept is the fact that each distinct excitase configuration determines uniquely the response of an enzymatic neuron network to each input pattern. Consequently, each network structure corresponds to an excitase configuration of a network of m k-compartment enzymatic neurons. We represent each structure as a binary vector $z = [z_{11}, z_{12}, \ldots, z_{1k}, z_{21}, \ldots, z_{mk}]$. Let $Z = \left\{ z_i \text{ in } \left\{ 0,1 \right\}^{km} \right\}$ denote the set of possible excitase configurations. The learning goal is to find a specific structure z_i in Z whose associated behavior corresponds to a satisfactory performance of the task being learned. The learning algorithm assigns a performance value to each structure tried. This performance value indicates how well a system represented by a given structure performs the task being learned. Thus, each task associates a metric space P with the space of structures Z. Let $B:Z \longrightarrow P$ be a mapping which assigns to each structure z_i in Z a performance p_i in P, $p_i = B(z_i)$, with respect to a specific task t. The mapping B formalizes the fact that a unique behavior, that yields performance p_i, is associated with each structure z_i in Z.

However, in general, it is not possible to describe in a closed mathematical form the behavior of the systems studied. This is because, in most cases, the combined effect of the characteristics of the systems subject to learning, the types of tasks involved, and the nature of the evolutionary learning process, all of which have an impact on the realization of this function, cannot be described analytically. For the mentioned reasons, we use computational modeling and computer simulation to determine the performance measures p_i corresponding to specific structures z_i, that is, to realize the function B.

The actual shape of this function is jointly determined by the type of enzymatic neurons involved and the nature of the specific task being learned. This can be expressed formally if we think of B as the composition of two functions U and V as follows: Let $U:Z \longrightarrow F$, where Z is the set of possible structures, and F is the set of functions representing the behaviors of structures z_i in Z. Let $V:F \longrightarrow P$ be a function that assigns a performance measure $p_i \, \varepsilon \, P$ to each function $f_i \, \varepsilon \, F$. Then, we have $B = V \times U$, with $p_i = VU(z_i)$. This is illustrated in Figure 12.3. The advantage of this formulation is that it makes it easier to relate each of the mappings U and V to their specific relevant factors.

Functions representing the behavior of the systems. Each $f_i \, \varepsilon \, F$ is a function $f_i:I \longrightarrow Y$, where I is the set of input patterns to which the enzymatic neuron networks may be subjected, and Y is the set of possible responses of the network represented by structure z_i to patterns in the input set I. In other words, each f_i represents the behavior of an enzymatic neuron with structure z_i. In the robot navigation task, for example, I is the set of sequences of input patterns a robot's control system receives in its route toward the target. Each such sequence, $T_i = <I(1), I(2), \ldots, I(N)>$, where N is the number of steps within which the robots must reach the target, represents successive inputs to the networks at each step during the performance of the task. Each $I(i) = <I_1, I_2, \ldots, I_m>$, I_j in [a,b], $j = 1, \ldots, m$, where a, b $\geq$ 0 are integers, represents input frequencies impinging on the neurons at step i of the execution of the task. In the case of discrete neurons, each input pattern is encoded as a binary vector, with $I_j \, \varepsilon \, [0,1]$. In simulations with continuous neurons, on the other hand, typical firing frequencies fall in the interval [0,10]. $R_{ik} = <O(1), O(2), \ldots, O(N)>$, $R_{ik} \, \varepsilon \, Y$, is a sequence of output patterns, the responses of the networks to input patterns impinging upon them at steps 1 through N. Each output pattern, $O(i) = <O_1, O_2, \ldots, O_k>$, where each $O_j \, \varepsilon \, [c,d]$, and c,d $\geq$ 0, represents the firing frequency of neuron j in response to input pattern $I(i)$. Again, for discrete neurons, each $O(i)$ is a binary vector, whereas in our simulations with continuous neurons firing frequencies in the output patterns were integers falling, typically, in the interval [0,6].

For a given task t, the function f_i realized by each structure z_i is determined by the type of neuron involved and its associated lower level dynamics. It is this dynamics which defines the time evolution of the excitation levels of the various compartments on the neuron's membrane and, therefore, its firing behavior. The upper level dynamics, on the other hand, is defined by the type of interaction assumed to exist between each neuron and its environment. It is therefore specified in terms of the encodings used to translate the state of the environment into input patterns and the interpretation of network responses in terms of effector actions.

The performance evaluation function. The task to be learned, including the encoding conventions for the input patterns and the interpretation of network responses in terms of effector actions, associates a metric space P with the set of functions, F. The performance evaluation function V assigns a unique performance value p_i in P to each function f_i in F, determined by structure z_i in Z, which measures the degree of success with which systems realizing such functions perform the task being learned. In the robot navigation task, for example, the performance evaluation function used is the euclidean distance between the final position of the robots in the grid and the position of the target, typically the origin of the coordinate system. In the simulations, however, the performance of the function f_i is only partially evaluated.

The role of simulation. As mentioned earlier, the functions in F cannot, in general, be represented in closed mathematical form. However, many interesting features of evolutionary learning processess can be identified and studied following the constructive approach of computational modeling and simulation. Earlier simulation experiments on evolutionary learning with enzymatic neurons have focused mainly on the determination of the types of task which are amenable to evolutionary learning; the identification, characterization, and analysis of the parameters of the learning algorithm (Kampfner and Conrad 1983a); and on the study of intraneuronal information processing within the framework of double dynamics (Kirby and Conrad 1984, 1985). Our purpose here is to highlight features of evolutionary learning processes which have become apparent through careful observation of the interplay between the evolutionary mechanisms implemented and the structure-behavior relationship of enzymatic neuron-based systems. The roles of gradualism and generalization, which are central in the characterization of this interplay, are discussed in the following sections.

Evolutionary Learning with Discrete Neurons

In an extensive number of simulation experiments on evolutionary learning with discrete neurons (Kampfner 1981; Kampfner and Conrad 1983a), it was found that the number of runs (evolutionary cycles with the learning algorithm) needed to learn to dichotomize patterns of n binary digits appears to grow only polynomially with problem size (i.e., with 2^n, the number of input patterns of n binary digits). Although this result was not tested for a statistically significant number of cases, it became evident that evolutionary learning is significantly better than pure random search. It is important to notice that for this kind of task, when random search is used, the time complexity of the learning process grows exponentially with problem size.

The Role of Gradualism

By gradualism we mean the ability of systems to show a slight change in their function concomitant with a correspondingly slight change in their structure. Discrete neurons learning pattern recognition tasks have the gradualism property if a small change in their excitase configuration results in a correspondingly small change in function. A step-by-step change in the structures is one of the features incorporated in the learning algorithm that ensures that this requirement is met by the simulated systems. The results of the simulations indicate that the tasks in which learning was accomplished efficiently were those in which gradualism was present. On the other hand, in the absence of gradualism, a significant slowdown, and in some cases, a complete stagnation of the learning process, was observed. Pattern recognition with arbitrary linear neurons (that is, cases in which each excitase enables a neuron to respond to an arbitrary subset of input patterns instead of a single one) is an example in this case. Similar results were obtained in experiments involving learning patterns of sequential behavior of networks of enzymatic neurons with feedback (Kampfner and Conrad 1983b).

Evolutionary Generalization

Evolutionary learning can be usefully analyzed from the point of view of the generalization concept previously defined. From this viewpoint, we look at evolutionary learning as a process in which structures associated with increasingly better performance are successively selected as the basis for the generation of new structures, each potentially capable of yielding even better performance, as the evolutionary process

progresses. The gradual, step-by-step, modification of the behavior of
the learning systems through the evolutionary learning process also
implies that a great many features of these systems are preserved from
one evolutionary cycle to the next. Furthermore, features which are
useful in most—or all—the environmental states to which a system
will be exposed are, clearly, likely to be preserved in the evolving
systems. Now, if the features developed up to some point in the
evolutionary process enable the systems to cope with a broad range
of environmental states, some of which the systems haven't yet been
exposed to, they will also enable the systems to cope with these yet
"unseen" environments. To the extent that this capability of anticipation
of environmental features exists, the evolving systems are said to
generalize on the basis of their current "knowledge." We will refer to
this kind of generalization, which stems from the nature of the evolutionary process, as evolutionary generalization.

In the context of certain tasks, generalization learning can be described as a rule-learning process. Here, learning to produce an adequate
response to patterns in a given class amounts to learning a rule implicit
in the performance of the task concerned. Clearly, if the response
learned applies also to patterns not included in the training set, the
rule-learning process amounts to generalization (that is, the learning
systems are capable of applying the rule in response to unseen patterns
of the class to which the rule applies). In the robot navigation task,
an implicit rule might be, for example, "counteract the random forces
acting on the robots at each time step." This is, of course, a statement
of the rule as it would appear to an external observer. Robots learning
such a rule would simply develop an adequate response to the class
of input patterns associated with that rule. Of course, such a response
is developed on the basis of an appropriate excitase configuration.
Thus, learning a rule amounts to developing an excitase configuration
that produces a behavior that corresponds to the application of that
rule. An appropriate action in the context of this rule would be, for
instance, for a robot to move to the South whenever the forces were
pushing it to the North (assuming that the target is located to the
South).

Observations from simulation experiments with discrete neurons
appear to be consistent with this kind of generalization. In the robot
navigation task, for example, it was observed that robots first learned
to handle certain aspects of the environment that were advantageous
from the viewpoint of the characteristics of the task, rather than from
the viewpoint of a specific path to be followed in order to successfully
perform the task. As more traits were successively incorporated into
the system's repertoire of behaviors, the ability of the systems to

perform effectively was also increased until, finally, the task could be performed in a satisfactory way.

Generalization can also be said to increase the efficiency of evolutionary processes. As a prerequisite for evolutionary adaptation, gradualism is essential for evolutionary generalization, since it requires that a sequence of excitase configurations, associated with a monotonic increase in performance, be developed. Given such a sequence, generalization then speeds up learning to the extent that rules associated with excitase configurations which appear earlier in the sequence also relate to excitase configurations required later. An important advantage of the evolutionary generalization capability can be seen here in terms of the ability gradually developed by the learning systems to respond successfully to environmental features present in situations not already experienced. Clearly, this type of generalization can be seen as a means of speeding up the learning process in evolutionary systems.

Evolutionary generalization can be characterized formally as follows. Let p be the required performance for a given task t. Let z_i be a structure in Z and T_G be some set of input sequences. If a system S_i (with structure z_i) yields performance $p_{ij} = V(f_i(T_j)) = p$, for some input sequence $T_j \, \varepsilon \, T_G$, then S_i is said to generalize over T_G to the extent to which $p_{ik} = V(f_i(T_k))$ is close to p, for each T_k in T_G. Clearly, if T_G is some training set, and S_i generalizes over T_G after having been exposed through learning only to T_s, T_s in T_G, the learning process is necessarily speeded up. Furthermore, the generalization capability guarantees that a task can be learned with a training set (that is, without having to expose the system to all possible input patterns).

Evolutionary Learning with Continuous Neurons

In continuous neurons, above-threshold concentrations of cyclic nucleotides that activate excitase enzymes—thus causing the neurons to fire—are the result of continuous diffusion processes taking place on the cell's membrane. As a result, in continuous neurons, a single excitase enzyme can be activated by a number of input patterns instead of a single one. By virtue of this dynamics, continuous neurons have intrinsic generalization capabilities. However, this dynamics also imposes constraints on these systems in terms of their possible behaviors and the way these behaviors can evolve. The impact of continuous dynamics on evolutionary learning is discussed in the following sections. The focus is on conditions which make intrinsic generalization more effective.

Effect of Continuous Dynamics on Problem Size

The richer range of behaviors that continuous neurons exhibit through the interplay between upper and lower level dynamics endows them with more powerful pattern recognition capabilities. The simulation of continuous dynamics makes it possible to analyze a richer range of behaviors of enzymatic neuron-based systems in a natural way. Specifically, the behavior of networks of continuous neurons can be simulated in terms of temporal, as well as spatial, patterns of firing. A continuous neuron with 8 dendritic inputs and a modest range of input frequencies, encoded as integers in the interval [0,10], can be exposed to as many as 8^{11} distinct input patterns. For a discrete neuron with the same number of dendritic inputs, the corresponding input range is only 2^8. In simulations of the robot navigation task with continuous neurons, for example, the robot's control system consisted of a single-layer network of four 11-excitase neurons. As mentioned in the section on evolutionary learning, robots with continuous neurons can detect the magnitude of the random forces and their distance to the target in terms of firing frequencies of the inputs. The number of distinct environmental states that can be encoded as input patterns of this kind (including direction of the forces and relative position with respect to the target) is approximately $3^4 \times 7^4 \simeq 2^{20} \simeq 160,000$. This means that we would need, in the case of discrete neurons, at least 20-bit patterns to encode the inputs, and at least 2^{20} distinct excitase enzymes to choose from in order to be able to learn and perform a comparable task.

The range of possible behaviors of the systems depends also on the number of information-processing elements required to perform a given task. In the robot navigation task, integer output frequencies in the range [0,4] were considered. This allows for 625 distinct responses in a network of four continuous neurons. With binary inputs, we would need at least 10 neurons to encode a comparable number of distinct behaviors. Thus, in the discrete neuron case, we would need a single-layer network of at least ten 256-excitase neurons, each processing eight 20-bit patterns to perform a task comparable to that performed by a network of four 11-excitase continuous neurons, each processing an 8-element integer vector. This shows a dramatic reduction of the size of the search space (that is, the number of possible excitase configurations of the networks), with a correspondingly dramatic reduction of the complexity of the learning process. This reduction is subject to the constraints imposed by continuous dynamics on the types of tasks that these systems can perform. The size of the search space, however, is still considerably large for most cases of interest.

Intrinsic Generalization, Constraints, and Gradualism

The intrinsic generalization capability of continuous neurons makes possible a speedup of the learning process in addition to that allowed by evolutionary generalization alone. However, it also imposes constraints on the class of behaviors that are amenable to evolutionary learning—that is, behaviors consistent with the gradualism property. In this respect, it seems intuitively clear that the greater the intrinsic generalization capability of the neurons (i.e., the larger the number of patterns added to, or deleted from, the "recognition set" of an enzymatic neuron in any single evolutionary cycle), the less the number of tasks for which a sequence of structures with the required properties exists. When a sequence with these properties exists for a given task, we say that the task is consistent with the intrinsic generalization properties of the neurons involved. Although continuous dynamics was shown to reduce considerably the size of the problem, we suggest here that intrinsic generalization is still crucial for a sufficient speedup of the learning process. Obviously, an important aspect of task consistency is the structure of the task itself. In this respect, a useful characterization of tasks structures is in terms of the hierarchical ordering they establish on the classes of input patterns. In this hierarchy, input pattern classes associated with more general features of the environment would be in a higher hierarchical level. Evolutionary learning would then be faster if responses to more general patterns are learned earlier in the process. This also implies an ordering in the set of rules defining a task, where rules associated with responses to more general aspects of the environment would be higher in the rules hierarchy. Simulation experiments involving the robot navigation task with continuous neurons show results consistent with intrinsic generalization as defined earlier. For instance, in the first stages of the learning process, the robots showed some kind of "gliding" behavior. They let themselves be pushed by "favorable" random forces. This kind of behavior gave them selective advantage further enhanced later by their ability to take steps in the right direction. A third component of the learned behavior involved adjusting the intensity of their responses to the intensity of the environmental inputs. In the case of inputs where frequency of firing encoded the distance of the robots from the target, the requirement is that the response to inputs of higher frequency (associated with a larger distance) be correspondingly stronger (meaning an action involving a greater displacement on the grid).

This necessarily informal description suggests that the robot navigation task is consistent with the dynamics used for the simulation of continuous neurons. It also suggests that the critical speedup of the learning process obtained was caused by both evolutionary and intrinsic generalization. In the experiments described, the task was learned in at most 28 evolutionary cycles. Of course, additional experiments and a more rigorous analysis are required for a more accurate characterization of the role of intrinsic generalization in evolutionary learning.

Summary and Conclusions

This chapter presents a view of evolutionary learning with enzymatic neuron-based systems that emphasizes their structure-function relationship. The evolutionary learning process is seen as a search on the space of structures that define the behavior of the learning systems. We studied evolutionary learning in systems with discrete and continuous neurons. Both types of systems incorporate the double dynamics principle as the underlying basis of their behavior. Although both discrete and continuous neurons have essentially the same upper level dynamics, their lower level dynamics show important differences in the size and topology of the search space, and the problem domains to which each can be applied. The basic requirement for effective evolutionary adaptation is gradualism, which implies that there exist sequences of structures, each showing a monotonic increase in their performance values.

Discrete neurons are found to be an adequate substrate for evolutionary learning involving unconstrained tasks. Our simulation experiments show that gradualism enhances the evolutionary learning capabilities of systems based on discrete neurons performing pattern recognition/pattern generation tasks. There is a significant improvement in the speed of evolutionary learning as compared with random search, for which the time required to learn the task is an exponential function of problem size. With evolutionary learning, the time complexity of the learning process appears to grow only polynomially with problem size, at least for the cases empirically determined in our simulation experiments.

The concept of evolutionary generalization is suggested as an inherent feature of evolutionary learning processes that partly explains this speedup. The degree of generalization that can be achieved by systems subject to evolutionary learning depends also on the task structure. We characterize this structure in terms of the hierarchy it establishes on the classes of input patterns to which the learning systems are exposed. Essentially, each class of patterns represents an equivalence class in the sense that patterns in each class require similar responses from the systems for the adequate performance of the task. This hierarchy reflects the generality of the environmental features encoded by the patterns. If a given task has an associated environment with classes of high generality, evolutionary generalization is likely to produce a great speedup of the learning process.

Continuous neurons allow for a significant reduction in the number of information-processing elements required (number of distinct types of excitase enzymes), as compared with discrete neurons performing comparable tasks. In the experiments reported, the reduction is estimated to be from 12 discrete neurons of 256 excitases each, to only four continuous neurons of 11 excitases. The corresponding reduction of the search space is from $2^{256 \times 12}$ to $2^{11 \times 4}$ elements. Continuous neurons also show the intrinsic generalization capability. We believe that intrinsic generalization, as a means of increasing the efficiency of the learning process, is still critical. Although continuous dynamics may allow for a considerable reduction of the size of the search space, this space is still very large in most cases of interest. However, this increase in efficiency, and, in fact, the amenability of the systems to evolutionary adaptation, is subject to constraints imposed by the dynamics on their possible behaviors. These constraints can be related to evolutionary processes in terms of their impact on the susceptibility of the structures to gradual evolution. This view provides a criterion for consistency of the dynamics with the task to be learned by the systems.

Thus, the interplay between the structure-function relationship of adaptive systems and the mechanisms involved in evolutionary adaptation can be usefully related to gradualism and the generalization capabilities of the learning systems. For highly constrained systems, if the task is consistent with the structure-function relationship of the learning system, intrinsic generalization allows for a speedup of the learning process in addition to that provided by evolutionary generalization alone.

The combination of elements of continuity and gradual modifiability with decision-making capabilities of systems is, indeed, the main contribution of the double dynamics to the efficiency-adaptability of evolutionary systems (Conrad 1985). The cost is paid in terms of a loss of effective programmability, as Conrad's trade-off principle also asserts. Our simulation experiments illustrate this principle by showing that efficiency and adaptability of information processing are paid with constraints on the type of tasks that can be learned, hence, universality of the information processing primitives used (i.e., effective programmability). A problem for further study is the precise characterization of tasks that can be learned with continuous dynamics and how this dynamics could itself evolve.

Bibliography

Conrad, M., 1974: Evolutionary learning circuits, *Journal of Theoretical Biology* 46:167–188.

Conrad, M., 1976: Complementary molecular models of learning and memory, *Biosystems* 8:119–138.

Conrad, M., 1985: On design principles for a molecular computer, *Communications of the ACM* 28:5.

Kampfner, R., 1981: *Computational modeling of evolutionary learning*, Ph.D. thesis, University of Michigan, Ann Arbor.

Kampfner, R., and M. Conrad, 1983a: Computational modeling of evolutionary learning processes in the brain, *Bulletin of Mathematical Biology* 45:931–968.

Kampfner, R., and M. Conrad, 1983b: Sequential behavior and stability properties of enzymatic neuron networks, *Bulletin of Mathematical Biology* 45:969–980.

Kirby, K.G., and M. Conrad, 1984: The enzymatic neuron as a reaction-diffusion network of cyclic nucleotides, *Bulletin of Mathematical Biology* 46:765–782.

Kirby, K.G., and M. Conrad, 1985: Intraneuronal dynamics as a substrate for evolutionary learning, *Physica D.*

Winston, P., 1975: Learning structural descriptions from examples, in P. Winston (ed.), *The Psychology of Computer Vision*, McGraw-Hill, New York.

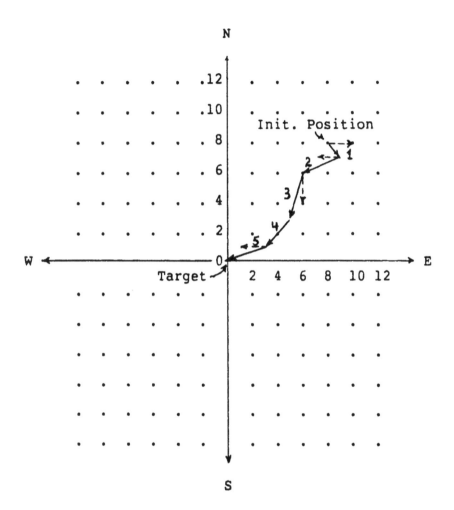

FIGURE 12.1 Route followed by a simulated robot that reached
the target in the five steps allowed. In this experiment, the
robot performed the task correctly after 28 evolutionary cycles.
In the figure, broken arrows represent direction and magnitude
of random forces. Solid arrows represent the displacement of the
robot at each step in the execution of the task. Numbers by the
arrows indicate the step in the execution of the task at which
the corresponding move was made.

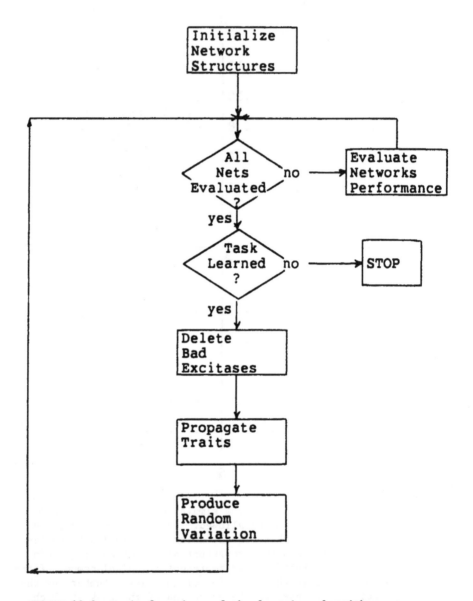

FIGURE 12.2 Basic functions of the learning algorithm

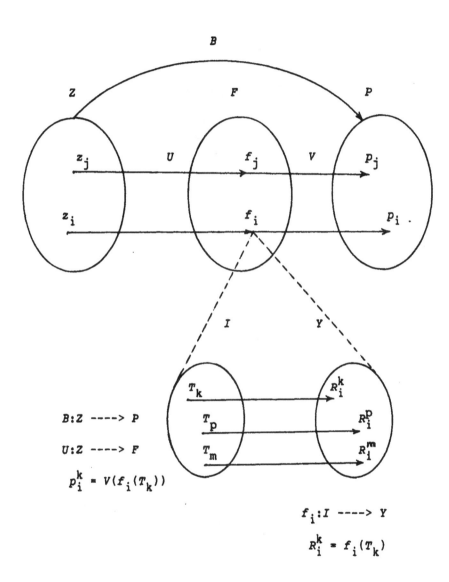

FIGURE 12.3 Relationship between the structure of learning
systems and the performance associated with their function

13. The Application of Algorithmic Probability to Problems in Artificial Intelligence

Introduction

This chapter covers two topics. First, it provides an introduction to algorithmic complexity theory: how it defines probability, some of its characteristic properties, and past successful applications. Second, it reviews the theory's applications to artificial intelligence (AI), where it promises to give near-optimum search procedures for two very broad classes of problems.

Algorithmic probability of a string s is the probability of that particular string being produced as the output of a reference universal Turing machine with random input. It is approximately $2^{-l(p)}$, where l(p) is the length of the shortest program p that will produce s as output. l(p) is the Kolmogorov complexity of s—one form of algorithmic complexity (Kolmogorov 1965, 1968).

Algorithmic complexity has been applied to many areas of science and mathematics. It has given us a very good understanding of randomness (Martin Lof 1966). It has been used to generalize information theory (Chaitin 1974), and has clarified important concepts in the foundations of mathematics (Chaitin 1975a). It has given us our first truly complete definition of probability (Solomonoff 1960, 1964, 1978).

The completeness property of algorithmic probability means that if there is any describable regularity in a body of data, our system is guaranteed to discover it using a relatively small sample of the data. It is the only probability evaluation method known to be complete. As a necessary consequence of its completeness, this kind of probability must be incomputable. Conversely, any computable probability measure cannot be complete.

Can we use this incomputable probability measure to obtain solutions to practical problems? A large step in this direction was Levin's search procedure that obtains a solution to any P or NP problem within a constant factor of optimum time. The "constant factor" may, in general, be quite large. Under certain reasonable conditions and for a broader class of problems than Levin originally considered, the "constant factor" must be less than about 4.

The P or NP class originally considered contained machine inversion problems: we are given a string s and a machine M that maps strings to strings. We must find in minimum time a string x such that $M(x)$ = s. Algebraic equation solving, symbolic integration, and theorem proving are examples of this broad class of problems.

However, Levin's search procedure also applies to another broad class of problems: time-limited optimization problems. Given a time limit T, and a machine M that maps strings to real numbers, to find within time T the string x such that $M(x)$ is as large as possible. Many engineering problems are of this sort: for example, designing an automobile in 6 months satisfying certain specifications having minimum cost. Constructing the best possible probability distribution or physical theory from empirical data in limited time is also of this form.

In solving either machine inversion problems or time-limited optimization problems, it is usually possible to express the information needed to solve the problem (either heuristic information or problem-specific information) by means of a conditional probability distribution. This distribution relates the string that describes the problem to any string that is a candidate solution to the problem. If all of the information needed to solve the problem is in this probability distribution, and we do not modify the distribution during the search, then Levin's search technique is within a factor of 2 of optimum. If we are allowed to modify the probability distribution during the search on the basis of our experience in attempting to solve the problem, then Levin's technique is within a factor of 4 of optimum.

The efficacy of this problem-solving technique hinges on our ability to represent all of our relevant knowledge in a probability distribution. To what extent is this possible? For one broad area of knowledge this is certainly easy to do: this is the kind of inductive knowledge obtained from a set of examples of correctly worked problems. Algorithmic probability obtained from examples of this sort is in just the right form for application of our general problem-solving system. Furthermore, when we have other kinds of information that we want to express as a probability distribution, we can usually hypothesize a sequence of examples that would lead to the learning of that information by a human. We can then give that set of examples to our induction system, and it will acquire the same information in appropriate probabilistic form.

While it is possible to put most kinds of information into probabilistic form using this technique, a person can, with some experience, learn to bypass this process and express the desired information directly in probabilistic form. We will show how this can be done for certain heuristic techniques, such as planning, analogy, clustering, and frame theory.

The use of a probability distribution to represent knowledge not only simplifies the solution of problems, but it also enables us to put information from many different kinds of problem-solving systems into a common format. Then, using techniques that are fundamental to algorithmic complexity theory, we can compress this heterogeneous mass of information into a more compact, unified form. This operation corresponds to Kepler's laws summarizing and compressing Tycho Brahe's empirical data on planetary motion. Algorithmic complexity theory has this ability to synthesize, to find general laws in masses of unorganized and partially organized knowledge. It is in this area that its greatest value for AI lies.

Algorithmic Complexity

The earliest application of algorithmic complexity was to devise a formal theory of inductive inference (Solomonoff 1960, 1964, 1978). All induction problems are equivalent to the problem of extrapolating a long sequence of symbols. Formally, we can do this extrapolation by Baye's theorem, if we are able to assign an a priori probability to any conceivable string of symbols, x.

This assignment can be made in the following manner: Let x be a string of n binary symbols. Let M be a universal Turing machine with 3 tapes: a unidirectional input tape, a unidirectional output tape, and an infinitely long bidirectional work tape. The unidirectionality of the input and output tapes assures us that if $M(s) = y$, then $M(ss') = yy'$, i.e., if s is the code of y, then if we extend s by several symbols, the output of M will be at least y and may possibly (although not necessarily) be followed by other symbols.

We assign an a priori probability to the string x, by repeatedly flipping a coin and giving the machine M an input 1 whenever we have "heads" or 0 whenever we have "tails." There is some probability $P_M(x)$ that this random binary input string will cause M to have as output a string whose first n bits are identical to x. When constructed in this manner with respect to universal Turing machine M, $P_M(x)$ becomes the celebrated universal a priori probability distribution.

Conceptually, it is easy to calculate $P_M(x)$. Suppose s_1, s_2, s_3 ... are all of the possible input strings to M that can produce x (at least) as output. Let s'_1, s'_2, s'_3 ... be a maximal subset of $[s_i]$ such that no s'_i can be formed by adding bits onto the end of some other s'_j. Thus, no s'_j can be the "prefix" of any other s'_i.

The probability of s'_i being produced by random coin tossing is just

$$2^{-l(s'_i)}$$

where l (s'_i) is the number of bits in s'_i.

Because of the prefix property, the s'_i are mutually exclusive events, and so the probability of x being produced by any of them is simply

$$\sum_i 2^{-l(s'_i)}$$

which is therefore the value of $P_M(x)$.

To do prediction with $P_M(x)$ is very simple. The probability that x will be followed by 1 rather than 0 is

$P_M(x1)/(P_M(x0) + P_M(x1))$.

How accurate are these probabilities?

Suppose that $P(a_{n+1} = 1 \mid a_1 a_2 a_3 \ldots a_n)$ is a conditional probability distribution for the $(n + 1)^{th}$ bit of a binary string, given the previous n bits, $a_1 a_2 a_3 \ldots a_n$. Let us further postulate that P is describable by machine M with a program b bits long.

Let $P_M (a_{n+1} = 1 \mid a_1 a_2 \ldots a_n)$ be the corresponding probability distribution based on P_M. Using P and a suitable source of randomness, we can generate a stochastic sequence $A = a_1 a_2 a_3 \ldots a_n$. Both P and P_M are able to assign probabilities to the occurrence of the symbol 1 at any point in the sequence A based on the previous symbols in A.

It has been shown (Solomonoff 1978, pp. 426–427) that the total expected squared error between P and P_M is given by

$$E\sum_{m=1}^{n} (P_M(a_{m+1} = 1 \mid a_1 a_2 \ldots a_m) - P(a_{m+1} = 1 \mid a_1 a_2 \ldots a_m))^2 < b \ln \sqrt{2}$$

The expected final value is with respect to probability distribution, P. This means that the expected value of the sum of the squares of the deviations of P_M from P is bounded by a constant.

This error is much less than that given by conventional statistics, which is proportional to the logarithm of n. The disparity occurs because P is describable by a finite string of symbols. Usually, statistical models have parameters with an infinite number of bits in them, and so the present analysis must be applied to them in somewhat modified form. The smallness of this error assures us that if we are given a stochastic sequence created by an unknown generator, we can use P_M to obtain the conditional probabilities of that generator with much accuracy.

Cover (1974) has shown (see also Solomonoff 1978, p. 425) that if P_M is made the basis of a universal gambling scheme, its yield will be extremely large. It is clear that P_M depends on just what universal machine M is used. However, if we use much data for our induction, then the probability values are relatively insensitive to choice of M. This will be true even if we include as data information not directly related to the probabilities we are calculating. P_M will yield the best probability values obtainable with the available information.

While P_M has many desirable properties, it cannot ever be used directly to obtain probability values. As a necessary consequence of its "completeness"—its ability to discover the regularities in any reasonable sample of data—P_M must be uncomputable. However, approximations to P_M are always possible, and we will later show how to obtain close to the best possible approximations with given computational resources.

One common way to obtain approximations of a probability distribution to extrapolate the string x is to obtain short codes for x. In general, short programs for the sequence x correspond to regularities in x. If x was a sequence of a million 1's, we could describe x in a few words and write a short program to generate it. If x was a random sequence with no regularities, then the shortest description of x would be x itself. Unfortunately, we can never know that a sequence is random. All we can ever know is that we have spent a lot of time looking for regularities in it, and have not found any. However, no matter how long we have looked, we can't be sure that we wouldn't find a regularity if we looked for 10 minutes more!

Any legitimate regularity in x can be used to write a shorter code for it. This makes it possible to give a clear criterion for success to a machine that is searching for regularities in a body of data. It is an adequate basis for the mechanization of inductive inference.

A General System for Solving Problems

The problems solvable by the system fall in two broad classes: machine inversion problems and time-limited optimization problems. In both, the problem itself, as well as the solution, can be represented by a finite string of symbols.

We will try to show that most, if not all, knowledge needed for problem solving can be expressed as a conditional probability distribution relating the problem string (condition) to the probability of various other strings being solutions. We shall be interested in probability distributions that list possible solutions, with their associated probability values in decreasing order of probability.

We will use algorithmic complexity theory to create a probability distribution of this sort. Then, considerations of computational complexity lead to a near-optimum method to search for solutions. We will discuss the advantages of this method of knowledge representation: how it leads to a method of unifying the Babel of disparate techniques used in various existing problem-solving systems.

Kinds of Problems That the System Can Solve

Almost all problems in science and mathematics can be well approximated or expressed exactly as either machine inversion problems or time-limited optimization problems.

Machine inversion problems include NP and P problems. They are problems of finding a number or other string of symbols satisfying certain specified constraints. For example, to solve $x + \sin x = 3$, we must find a string of symbols (i.e., a number x) that satisfies this equation. Problems of this sort can always be expressed in the form $M(x) = c$. Here, M is a computing machine with a known program that operates on the number x. The problem is to find an x such that the output of the program is c.

Symbolic integration is another example of machine inversion. For example, we might want the indefinite integral of xe^x. Suppose M is a computer program that operates on a string of symbols that represent an algebraic expression and obtains a new string of symbols representing the derivative of the input string. We want a string of symbols s such that $M(s) = xe^x$.

Finding proofs of theorems is also an inversion problem.

Let Th be a string of symbols that represents a theorem.

Let Pr be a string of symbols that represents a possible proof of theorem Th.

Let M be a program that examines Th and Pr. If Pr is a legal proof of Th, then its output is "Yes," otherwise, it is "No."

The problem of finding a proof becomes that of finding a string s such that $M(Th,s) = $ Yes.

Many other problems can be expressed as machine inversion problems.

Another broad class of problems are time-limited optimization problems. Suppose we have a known program M that operates on a number or string of symbols and produces as output a real number between

0 and 1. The problem is to find an input that gives the largest possible output, and we are given a fixed time T in which to do this.

Many engineering problems are of this sort. For example, consider the problem of designing a rocketship satisfying certain specifications, having minimal cost, within the time limit of 5 years. Another broad class of optimization problems are induction problems. An example is to devise the best possible set of physical laws to explain a certain set of data, and doing this within a certain restricted time.

It should be noted that devising a good functional form for M—the criterion for how well the theory fits the data—is not in itself part of the optimization problem. A good functional form can, however, be obtained from algorithmic complexity theory.

The problem of extrapolating time series involves optimizing the form of prediction function, and so it, too, can be regarded as an optimization problem.

Another form of induction is "operator induction." Here, we are given an unordered sequence of ordered pairs of objects such as (1,1), (7,49), (−3,9). The problem is to find a simple functional form relating the first element of each part (the "input") to the second element (the "output"). In the example given, the optimum is easy to find, but if the functional form is not simple and noise is added to the output, the problems can be quite difficult. Some "analogy" problems on I.Q. tests are forms of operator induction.

In the most general kind of induction, the permissible form of the prediction function is very general, and it is impossible to know if any particular function is the best possible, only that it is the best found thus far. In such cases, the unconstrained optimization problem is undefined, and including a time limit constraint is one useful way to give an exact definition to the problem.

All of these constrained optimization problems are of the form: Given a program M, to find string x in time T such that $M(x) =$ maximum. In the examples given, we always knew what the program M was. However, in some cases, M may be a "black box," and we are allowed only to make trial inputs and remember the resultant outputs. In other forms of the optimization problem, M may be time varying and/or have a randomly varying component.

We will discuss here only the case in which the nature of M is known and is constant in time. Our methods of solution are, however, applicable to certain of the other cases.

In both inversion and optimization problems, the problem itself is represented as a string of symbols. For inversion problem $M(x) = c$, this will consist of the program M followed by the string, c.

For optimization problem $M(x) = max$, in time T, our problem is represented by the program M followed by the number, T.

For inversion problems, the solution x will always be a string of the required form.

For an optimization problem, a solution will be a program that looks at the program M, the time T, and results of previous trials, and from these creates as output the next trial input to M. This program is always representable as a string, just as is the solution to an inversion problem.

Before telling how to solve these two broad categories of problems, I want to introduce a simple theorem in probability.

At a certain gambling house, there is a set of possible bets available, all with the same big prize. The ith possible bet has probability p_i of winning, and it costs d_i dollars to make the ith bet. All probabilities are independent, and one can't make any particular bet more than once. The p_i need not be normalized.

If all the d_i are one dollar, the best bet is clearly the one of maximum p_i. If one doesn't win on that bet, try the one of next largest p_i, and so on. This strategy gives the least number of expected bets before winning.

If the d_i are not all the same, the best bet is that for which p_i/d_i is maximum. This gives the greatest win probability per dollar.

Theorem I: If one continues to select subsequent bets on the basis of maximum p_i/d_i, the expected total money spent before winning will be minimal.

In another context, if the cost of each bet is not dollars, but time t_i, then the betting criterion p_i/t_i gives least expected time to win.

In order to use this theorem to solve a problem, we would like to have the functional form of our conditional probability distribution suitably tailored to that problem. If it were an inversion problem, defined by M and c, we would like a function with M and c as inputs that gave us as output a sequence of candidate strings in order of decreasing p_i/t_i. Here, p_i is the probability that the candidate will solve the problem, and t_i is the time it takes to generate and test that candidate. If we had such a function, and it contained all of the information we had about solving the problem, an optimum search for the solution would simply involve testing the candidates in the order given by this distribution. Unfortunately, we rarely have such a distribution available, but we can obtain something like it from which we can get good solutions.

One such form has input M and c as before, but as output, it has a sequence of string probability pairs (a_1, p_1), (a_2, p_2). . . . p_i is the probability that a_i is a solution, and the pairs are emitted in order of decreasing p_i.

When algorithmic complexity is used to generate probability distributions, these distributions have approximately this form. How can we use such a distribution to solve problems? First, we select a small time limit T_0 and we do an exhaustive search of all candidate solution strings a_i such that $t_i/p_i < T_0$. Here, t_i is the time needed to generate and test a_i. If we find no solution, we double T_0 and go through the exhaustive testing again. The process of doubling T_0 and searching is repeated until a solution is found. The entire process is approximately equivalent to testing the a_i's in order of increasing t_i/p_i.

It's not difficult to prove Theorem II.

Theorem II: If a correct solution to the problem is assigned a probability p_j by the distribution, and it takes time t_j to generate and test that solution, then the algorithm described will take a total search time of less than $2\, t_j/p_j$ to find that solution.

Theorem III: If all of the information we have to solve the problem is in the probability distribution, and the only information we have about the time needed to generate and test a candidate is by experiment, then this search method is within a factor of 2 of the fastest way to solve the problem.

Levin (1973) used an algorithm much like this one to solve the same kinds of problems (see also Solomonoff 1984), but he did not postulate that all of the information needed to solve the problems was in the equivalent of the probability distribution. Lacking this strong postulate, his conclusion was weaker (i.e., that method was within a constant factor of optimum). Although it was clear that this factor could often be very large, he conjectured that, under certain circumstances, it would be small. Theorem III gives one condition under which the factor is 2.

In artificial intelligence research, problem-solving techniques optimum within a factor of 2 are normally regarded as much more than adequate, so a superficial reading of Theorem III might regard it as a claim to have solved most of the problems of AI! This would be an inaccurate interpretation.

Theorem III postulates that we put all of the needed problem-solving information (both general heuristic information and problem-specific information) in the probability distribution. To do this, we usually use the problem-solving techniques that are used in other kinds of problem-solving systems, and translate them into modifications of the probability distribution. The process is analogous to the design of an expert system by translating the knowledge of a human expert into a set of rules. While I have developed some standard techniques for doing this, the translation process is not always a simple routine. Often, it gives the benefit of viewing a problem from a different perspective, yielding new, better understanding of it. Usually, it is possible to simplify and improve problem-solving techniques a great deal by adding probabilistic information.

The Overall Operation of the System

We start with a probability distribution in which we have placed all of the information needed to solve the problems. This includes both general heuristic as well as problem-specific information. We also have the standard search algorithm of Theorem II which has been described.

The first problem is then given to the system. It uses the search algorithm and probability distribution to solve the problem. (See Figure 13.1.) If the problem is not solvable in acceptable time, then the t_j/p_j of the solution must be too large. t_j can be reduced by using a faster computer, or by dividing up the search space between several computers, or by using faster algorithms for various calculations. p_j can be increased by assigning short codes (equivalent to high probabilities) to commonly used sequences of operations. Solomonoff (1964) tells how to do this on pages 232–240.

After the first problem is solved, we have more information than we had before (i.e., the solution to the first problem), and we want to put this information into the probability distribution. This is done by a process of "compression." We start out with the probability distribution that has been obtained by finding one or more short codes for a body of data, D. Let the single length equivalent of all of these codes be L_D. Suppose the description length of our new problem solution pair PS is L_{PS}. "Compression" consists of finding a code for the compound object (D, PS) that is of length less than $L_D + L_{PS}$. Compression is expressible as a time-limited optimization problem. If the system took time T to solve the original problem, we will give it approximately time T for compressing the solution to this problem into the probability distribution. This compression process amounts to "updating" the probability distribution. (See Figure 13.2.)

There may seem to be some logical difficulty in having a machine work on the improvement of a description of part of itself. However, we need not tell the machine that this is what it is doing. In the system described, there is no way for it to obtain or use such information.

After compression, we give it the next external problem to solve, followed by another updating or compression session. If the probability distribution contains little information relevant to compression, then the compression can be done by the system's (human) operator. Eventually, the distribution will acquire enough information (through problem-solving experience or through direct modification of the probability distribution by the operator) to be able to do useful compression in the available time without human guidance.

What are the principal advantages of expressing all of our knowledge in probabilistic form?

1. We have a near optimum method of solving problems when the knowledge needed to solve them is in that form.
2. It is usually not difficult to put our information into that form, and when we do so, we often find that we can improve problem-solving methods considerably. Traditionally, a large fraction of AI workers have avoided probabilistic methods. By ignoring probability data, they are simply throwing away relevant information. Using it in an optimum manner for search can do nothing less than speed up the search process and give a more complete understanding of it.
3. Theorem II gives us a unique debugging tool. If the system cannot find a known solution to a problem in acceptable time, analysis of the t_j and p_j involved will give us alternative ways to reduce $2\ t_j/p_j$.
4. Once our information is in this common format, we can compress it. The process of compressing information involves finding shorter codes of that information. This is done by finding regularities in it, and expressing these regularities as short codes.

Why do we want to compress the information in our probability distribution?

1. By compressing it, we find general laws in the data. These laws automatically interpolate, extrapolate, and smooth the data.
2. By expressing a lot of information as a much smaller collection of laws, we are in a far better position to find higher order laws than we would be if we worked with the data directly. For example, Newton's laws were much easier to discover as an outgrowth of Kepler's laws than it would have been for Newton to derive them directly from purely experimental data.
3. The compressed form of data is easier for humans to understand, so they may better know how to debug and improve the system.
4. The processes of interpolating and extrapolating problem-solving methods automatically create new trial problem-solving methods

that have high probability of working. This makes it possible for the system to go beyond the insular techniques originally built into it, and gives us the closest thing to true creativity that we can expect in a mechanized device.

In short, compression of information in the probability distribution transforms our system from a collection of disconnected problem-solving techniques into a unified, understandable system. It promises to make AI an integrated science rather than a compendium of individually developed, isolated methodologies.

Using Probability Distributions to Represent Knowledge

A critical question in the foregoing discussion is whether we can represent all of the information we need to solve our problems through a suitable probability distribution. I will not try to prove this can always be done, but will give some fairly general examples of how to do it.

The first example will be part of the problem of learning algebraic notation from examples. The examples are of the form

35, 41, + : 76
8, 9, × : 72
−8, 1, + : −7. . . .

The examples all use $=$, $-$, $\times$, and $\div$ only. The problem is for the machine to induce the relationship of the string to the right of the colon to the rest of the expression.

To do this, it has a vocabulary of 7 symbols: R_1, R_2, and R_3 represent the first 3 symbols of its input.

Add, Sub, Mul, Div represent internal operators that can operate on the contents of R_1, R_2, and R_3 if they are numbers.

The system tries to find a short sequence of these 7 symbols that represents a program expressing the symbol to the right of the colon in terms of the other symbols.

35, 41, + : 76 can be written as 35, 41, + :R_1, R_2, Add.

If all symbols have equal probability to start, the subsequence R_1, R_2, Add has probability $1/7 \times 1/7 \times 1/7 = 1/343$. If we assume 16-bit precision, each integer has probability $2^{-16} = 1/65536$, so $1/343$ is a great improvement over the original data.

We can code the right halves of the original expressions as

R_1, R_2, Add
R_1, R_2, Mul
R_1, R_2, Add

If there are many examples like these, it will be noted that the probability that the symbol in the first position is R_1 is close to one. Similarly, the probability that the second symbol is R_2 is close to one. This gives a probability of close to $1/7$ for R_1, R_2, Add.

We can increase the probability of our code further by noting that in expressions like

35, 41, = : R_1, R_2, Add,

the last symbol is closely correlated with the third symbol, so that knowing the third symbol, we can assign very high probability to the final symbols that actually occur.

I have spoken of assigning high and low probabilities to various symbols. How does this relate to length of codes? If a symbol has probability p, we can devise a binary code for that symbol (a Huffman code) having approximately $-\log p$ bits in it. If we use many parallel codes for a symbol (as in our definition of algorithmic probability), we can have an equivalent code length of exactly $-\log p$.

The second example is a common kind of planning heuristic. When a problem is received by the system, it goes to the "planner" module. The module examines the problem, and on the basis of this examination, assigns 4 probabilities to it: P_1, P_2, P_3, and P_4. (See Figure 13.3.)

P_1 is the probability that the quickest solution will be obtained by first breaking the problem into subproblems ("divide and conquer") that must all be solved. Module M_1 breaks up the problem, and sends the individual subproblems back to "planner."

P_2 is the probability that the quickest solution is obtainable by transforming the problem into several alternative equivalent problems. Module M_2 does these transformations, and sends the resultant problems to "planner."

P_3 is the probability of solution by method M_3. M_3 could, for example, be a routine to solve algebraic equations.

P_4 is the probability of solution by method M_4. M_4 could, for example, be a routine to perform symbolic integration.

The operation of the system in assigning probabilities to various possible solution trials looks much like the probability assignment process in a stochastic production grammar.

Because the outputs of M_1 and M_2 go back to "planner," we have a recursive system that can generate a search tree of infinite depth.

However, the longer solution trials have much less probability, and so when we use the optimum search algorithm, we will tend to search the high probability trials first (unless they take too long to generate and test).

The third example is an analysis of the concept "analogy" from the viewpoint of algorithmic probability.

The best-known AI work in this area is that of Evans (1963), a program to solve geometric analogy problems such as those used in intelligence tests.

We will use an example of this sort.

Given (a)△▽ (b)▽△ (c)⊥⊤ (d)♦♀ (e)♦♦ is d or e more likely to be a member of the set a, b, c?

We will devise short descriptions for the sets a, b, c, d and a, b, c, e, and show that a, b, c, d has a much shorter description.

The set a, b, c, d can be described by a single operator followed by a string of operands:
Op_1 = [print the operand, then invert operand and print it to the right].

A short description of a, b, c, d is then (1) Op_1 [description of △, description ▽, description of ⊥, description of ♦].

To describe a, b, c, e, we will also need the operators (Op : Op_2 = [print the operand, then print it again to the right].

A short description of a, b, c, e is then: (2) Op_1 [description of △, description of ▽, description of ⊥, Op_2 [description of ♦].

It is clear that (1) is a much shorter description than (2). We can make this analysis quantitative if we actually write out many descriptions of these two sets in some machine code.

If a, b, c, d is found to have descriptions of lengths 100, 103, 103, and 105, then its total probability will be

$$2^{-100} \times [1 + 2^{-3} + 2^{-3} + 2^{-5}] = 1.28135 \times 2^{-100}.$$

If a, b, c, e has description lengths 105, 107, 108, it will have a total probability of

$$2^{-100} \times [1 + 2^{-3} + 2^{-3} + 2^{-5}] = 2^{-100} \times .0429688.$$

The ratio of these probabilities is 29.8, so if these code lengths were correct, the symbols ♦♀ would have a probability of 29.8 times as great as ♦♦ of being a member of the set a, b, c.

The concept of analogy is pervasive in many forms in science and mathematics. Mathematical analogies between mechanical and electrical systems make it possible to predict accurately the behavior of one by analyzing the behavior of the other. In all of these systems, the pair of things that are analogous are obtainable by a common operator such as Op_1, operating on different operands. In all such cases, the kind of analysis that was used in our example can be directly applied.

The fourth example is a discussion of clustering as an inductive technique.

Suppose we have a large number of objects, and the ith object is characterized by k discrete or continuous parameters, $(a_{i1}, a_{i2}, \ldots a_{ik})$. In clustering theory, we ask, "Is there some natural way to break this set of objects into a bunch of 'clusters' (subsets) so that the elements within each cluster are relatively close to one another?"

Algorithmic probability theory regards this as a standard coding problem. The description of the set of objects consists first of a set of points in the space of parameters that corresponds to "centers of clusters." Each of the objects is then described by the name of its cluster, and a description of how its parameters differ from that of its cluster center. If the points of each cluster are distributed closely about their center, then we have achieved great compression of code by describing the cluster centers, followed by short codes giving the small distances of each point from its center.

The efficacy of this system depends critically on our formalism for describing the parameters. If we have any probabilistic information about the distribution of points, this can be used to define good metrics to describe the deviations of points from their respective centers.

The fifth example is a description of frame theory as a variety of clustering.

Minsky's (1975) introduction to frames treats them as a method of describing complex objects and storing them in memory. An example of a frame is a children's party. A children's party has many parameters that describe it: Who is giving the party? What time of day will it be given? Is it a birthday party? (i.e., Must we bring presents?) Will it be in a large room? What will we do? If we know nothing about it other than the fact that it's a party for children, each of the parameters will have a "default value"—this standard set of default parameters defines the "standard children's party." This standard can be regarded as a "center point" of a cluster space.

As we learn more about the party, we find out the true values of many of the parameters that had been given default assignments. This moves us away from the center of our cluster, with more complex (and hence, less probable) descriptions of the parameters of the party. Certain of the parameters of the party can, in turn, be described as frames having sets of default values which may or may not change as we gain more information.

Present State of Development of the System

How far have we gone toward realizing this system as a computer program? Very little has been programmed. The only program specifically written for this system is one that compresses a text consisting of a long sequence of symbols. It first assigns probabilities to the symbols. Then, new symbols are defined that represent short sequences appearing in the text (Solomonoff 1964, pp. 232–240), and they are also assigned probabilities. The principles of this program are very useful, since in most bodies of data, the main modes of compression are assignments of probabilities to symbols and defining new symbols.

We have studied compression of text by coding it as a stochastic, context-free grammar (Solomonoff 1964, pp. 240–253, 1975, pp. 276–277). Some work has been done on devising training sequences as a means of inserting information into the probability distribution (Solomonoff 1962, 1982).

It has been possible to take an existing AI system and "retrofit" it with the present probabilistic approach (Solomonoff 1975, pp. 276–277). Some of the best-known work on mechanized induction is Winston's (1975) program for learning structures from examples. It uses a training sequence of positive examples and close negative examples ("near misses"). After much computation, it is able to choose a (usually correct) structure corresponding to the examples.

The probabilistic form for this problem simplifies the solution considerably, so that probabilities for each possible structure can be obtained with very little calculation. The system is able to learn even if there are no negative examples, which is well beyond the capabilities of Winston's program. That probabilities are obtained rather than "best guesses" is an important improvement. This makes it possible to use the results to obtain optimum statistical decisions. "Best guesses" without probabilities are of only marginal value in statistical decision theory.

There are a few areas in which we haven't yet found very good ways to express information through probability distributions. Finding techniques to expand the expressive power of these distributions remains a direction of continued research.

However, the most important present task is to write programs demonstrating the problem-solving capabilities of the system in the many areas where representation of knowledge in the form of probability distributions is well understood.

Bibliography

Chaitin, G.J., 1974: Information-theoretic limitations of formal systems, *Journal of Automated Computing Machinery* 21:403–424, July.

Chaitin, G.J., 1975a: Randomness and mathematical proof, *Scientific American* 232:47–52, May.

Chaitin, G.J., 1975b: A theory of program size formally identical to information theory, *Journal of Automated Computing Machinery* 22:329–340, July.

Cover, T.M., 1974: *Universal Gambling Schemes and the Complexity Measures of Kolmogorov and Chaitin*, Report 12, Statistics Department, Stanford University, Calif.

Evans, T., 1963: A heuristic program for solving geometric analogy problems, Ph.D. dissertation, Massachusetts Institute of Technology, Cambridge, Mass.

Kolmogorov, A.N., 1965: Three approaches to the quantitative definition of information, *Information Transmission* 1:3–11.

Kolmogorov, A.N., 1968: Logical basis for information theory and probability theory, *IEEE Transactions on Information Theory* IT-14:662–664, September.

Levin, L.A., 1973: Universal search problems, *Problemy Peradaci Informacii* 9:115–116. Translated in *Problems of Information Transmission* 9:265–266.

Martin Lof, P., 1966: The definition of random sequences, *Information and Control* 9:602–619, December.

Minsky, M., 1975: A framework for representing knowledge, in P. Winston (ed.), *The Psychology of Computer Vision*, McGraw-Hill, New York.

Solomonoff, R.J., 1960: *A Preliminary Report on a General Theory of Inductive Inference*, ZTB-138, Zator Company, Cambridge, Mass.

Solomonoff, R.J., 1962: Training sequences for mechanized induction, in M. Yovits (ed.), *Self-Organizing Systems*, Spartan, Washington, D.C.

Solomonoff, R.J., 1964: A formal theory of inductive inference, *Information and Control* 1:1–22 and 2:224–254, March and June.

Solomonoff, R.J., 1975: Inductive inference theory—a unified approach to problems of pattern recognition and artificial intelligence, Fourth International Joint Conference on Artificial Intelligence, Tbilisi, Georgia, USSR, 274–280, September.

Solomonoff, R.J., 1978: Complexity-based induction systems—comparisons and convergence theorems, *IEEE Transactions on Information Theory* IT-24, 4:422–432.

Solomonoff, R.J., 1982: *Perfect Training Sequences and the Costs of Corruption— A Progress Report on Inductive Inference Research*, Oxbridge Research (P.O. Box 559) Cambridge, Mass., 02238.

Solomonoff, R.J., 1984: *Optimum Sequential Search*, Oxbridge Research (P.O. Box 559), Cambridge, Mass., 02238.

Winston, P., 1975: Learning structural descriptions from examples, in P. Winston (ed.), *The Psychology of Computer Vision*, McGraw-Hill, New York.

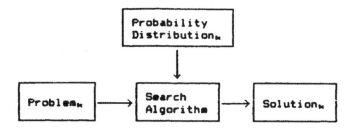

FIGURE 13.1   Problem solving using the search algorithm and probability distribution

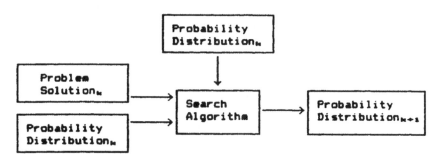

FIGURE 13.2   Problem solving via the compression process

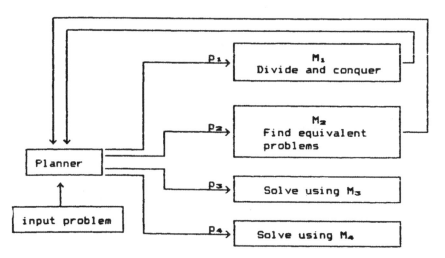

FIGURE 13.3   Problem solving through a probability assignment system

14. Novelty and Redundancy in Problem Solving

Introduction

The most highly developed metaphor of problem solving is that of search through a finitely branching network of alternatives (Newell and Simon 1972; Nilsson 1971; Lindsay et al. 1980). Human cognitive abilities doubtlessly encompass other important forms, but this paradigm clearly covers a wide variety of important cases. It has been studied sufficiently, both within the fields of psychology and, more recently and systematically, in artificial intelligence, that there is an emerging theoretical structure to it.

Formulation of problems in terms of a finitely branching network of alternatives—hereafter referred to as a *problem graph*—presumes a method of defining alternatives. Generally speaking, the theoretical work on the problem-graph paradigm has left this step exogenous to the theory. That is, it is assumed that the set of alternatives is given prior to the construction and exploration of the problem graph. For the human problem solver, however, the step of constructing the graph is usually nontrivial. One of the key issues is the definition and/or recognition of equivalence of alternatives. Our work addresses this issue, both from a theoretical point of view by attempting to provide formal definitions of equivalence, and from an empirical point of view by attempting to determine the relative ease with which human subjects can deal with different types of equivalence.

Problem Equivalence

To illustrate, we will first consider a familiar game, tic-tac-toe. A move consists of placing a mark symbolizing one of the two players in an unoccupied square of the 3 × 3 playing board, with the two players alternating. There are nine possible first moves, eight possible replies by the second player, and so forth. A straightforward problem graph analysis of this game represents it as a rooted tree, with nine branches from the root corresponding to the nine alternative moves open to the first player, eight from each of the nine new nodes thus generated, and so forth. Each node in the problem graph (here, a problem *tree*) has associated with it the game board configuration as it appears at that position. Thus, the root has a blank game board, the node at the end of the first blank has a blank game board save for the first player's symbol in, say, the top left corner, and so forth.

The problem-graph paradigm offers a number of methods for solving the problem of winning a tic-tac-toe game. One such is the minimax strategy, including numerous variations, such as alpha-beta (Newell, Shaw, and Simon 1958; Knuth and Moore 1975). Applying simple minimax to this game tree amounts to exhaustive search of the 9! = 362,880 terminal nodes[1] to determine which represent wins for the problem solver, and then backing these winning nodes up to the current node to select a move. This amount of search is well within the scope of contemporary computers, but not of contemporary humans.

When a child is first taught this game, he typically will not conceive of it in the problem-graph paradigm, unless he has had a great deal of experience with this sort of structured problem or has been explicitly instructed in the concept (and is old enough to understand it, generally not before 9 or 10 years of age). His first problem is to understand a host of prior concepts, such as what a game is, what a move is, what the markers represent, and so forth. The theories of problem-graph search offer no insights into how the child solves these pre-problems, and we will not address them either.

A second class of problems that the beginner must address includes learning what constitutes a win (three of his markers in a row), what constitutes a block, and what constitutes functionally different moves. While the experienced player finds these issues trivial, that is manifestly not the case with the beginner, especially a child. Simply learning to recognize each of the eight possible winning patterns generally requires fairly extensive practice; recognizing a near win (two in a row, of which there are 24 patterns) is even more difficult. These issues, too, are not addressed by problem-graph search theories. They are, however, a focus of our work.

Let us return to a more detailed analysis of tic-tac-toe to elaborate on this focus. In this game, a win is defined explicitly as "three identical markers in a row." This implicity defines a set of eight patterns representing wins; each pattern in turn may be instantiated by a larger number of actual board configurations by filling in the six remaining squares in a variety of ways. Thus, the eight "wins" previously alluded to are really eight equivalence classes of board configurations.

This notion of equivalence may be extended to include nonterminal configurations, in which case equivalence of configuration implies equivalence of move selection. For example, a board with an X in the center and an O in the top left corner has the same import for the purposes of *this* game as a board with an X in the center and an O in any other *corner*. For the initial configuration (a blank board), there are only three functionally distinct configurations; corner, center, or side square. An equivalence relation has partitioned the set of nine possible initial moves into three equivalence classes.

Clearly, the recognition of these various equivalences—in this instance, all of which derive from the symmetry of the game board—can vastly simplify the problem. The reduced game tree has only three branches from the root, and these in turn have only 2 or 3 alternatives from them. The reduced tree has fewer than 8,000 terminal nodes.[2] This brings it to within reach of exhaustive search by some people, and close for most people.[3]

It should be emphasized that the minimax strategy mentioned earlier is only one of many possible theories of search that might be applied in this context, yet all of them benefit greatly, although not equally, from the large reduction of the search graph achieved by the recognition of symmetries.

Next, we consider another example of strategy, a simple heuristic strategy often discovered and used by players of this game. This heuristic strategy does not guarantee a win even in those cases where the opponent makes a blunder, but it is nonetheless a better strategy by far than that used by most beginners. It consists of the following rules:

1. Make a winning move if you can.
2. If you cannot, block the opponent's potential win.
3. If he has none, move in the center if it is open.
4. If the center is occupied, move in a corner.
5. Otherwise, move in any open side position.

Note that this heuristic strategy basically incorporates *no more problem-solving knowledge than that inherent in the recognition of symmetries*. Yet, it is a viable strategy against most players.

Other problem-solving strategies that might be employed are to devise a merit factor for board positions (based on any of a variety of features) and to select the move that maximizes that factor; to memorize previous games and follow those that were wins; etc. In each of these cases, recognition of equivalence simplifies the strategy, often reducing it from unmanageable to manageable size. We will now consider other types of equivalence.

Specifying Problems and Equivalence Classes

We will now present an informal characterization of a class of problems, and show how this characterization can be used to determine certain kinds of equivalences. This will permit the development of algorithms for exploiting equivalence, and the design of experimental materials for the study of human ability to recognize redundancy and novelty in problem solving.

Our analysis will be restricted to those problems that can be stated in a *problem-specification language,* psl. A psl is a formally defined language that can be used to give precise specifications for a class of problems. These problem specifications are used to define equivalence classes of problems and solutions, but are not used to present problems to human problem solvers. A particular psl is a member of a class of languages that have context-free grammars and terminal symbols of the following kinds:

1. individual names,
2. typed variable names,
3. typed first-order predicates,
4. logical connectives *and, or, not,* and *implication,*
5. existential and universal quantifiers,
6. parentheses,
7. numerical predicates, such as $<, >, =,$
8. arithmetic operations, such as $+, -, \div, \times,$
9. a set of imperatives, such as *Find,* needed to specify goals,
10. specifications of the syntactic symmetries of the predicates.

Consider the following verbal problem statement from elementary algebra: *Jill can paint a wall in 5 hours, and Jack can paint the same wall in 7.5 hours. How long would it take for the two of them to paint the wall if they worked together?* This problem statement is in English, manifestly not a language meeting our specifications. However, the problem can with little effort be translated into a psl of the required sort. We introduce the predicate paintwall($\{$ S $\}$) to denote that members of the set S of named individuals can "paint the wall in question." Thus, paintwall($\{$ Jill $\}$) = 5 hours is the rendition of the first clause. The entire problem can then be stated: If paintwall($\{$ Jill $\}$) = 5 and paintwall($\{$ Jack $\}$) = 7.5, then Find paintwall($\{$ Jill,Jack $\}$). Here, by translating problems into a restricted psl, we have avoided the unsolved problem of English comprehension by computer.

We also wish to include some classes of problems that require simple diagrams for their expression. Note that in the definition of a psl, clause 3 permits the use of arbitrary first-order predicates. This vehicle will permit the coding of some pictorial information by use of such predicates as *left of, above, connected to,* and so forth. Thus, our analytic methods can be applied to these problems by making the appropriate translation. Obviously, not all game-tree problems can be formulated in psl. Nonetheless, a fairly wide range of nontrivial problems can be.

Equivalences Expressible in a psl

A psl permits variable names to be *typed,* that is, the range of permissible values to which a variable may be bound is specifiable. This permits classification of variables into such types as human personal names, real numbers, positive integers, signed integers, and sets of each of these types. The substitutability of individual names within these type restrictions defines several classes of equivalence-preserving transformations on problem statements that we will now informally describe.

Equivalence Under Substitutions of Individuals

In the preceding word problem, the names of the individuals, Jill and Jack, may be replaced by other names without altering the substance of the problem. Two such substitutions yield a problem that is in some important sense equivalent: *Alice can paint a wall in 5 hours, and Bill can paint the same wall in 7.5 hours. How long would it take Alice and Bill to paint the wall if they worked together?* An equivalence class of problems is generated in this way by substituting variables restricted to taking individual names as values for the specific names used in the initial statement of the problem: *M can paint a wall in 5 hours, and N can paint the same wall in 7.5 hours. How long would it take M and N to paint the wall if they worked together?* Conversely, this generically stated problem can lead to a set of specific problems by substituting specific values for the personal-name-valued variables, M and N.

Again, in our sample problem, the numerical values of work times can be replaced by other values. Two such substitutions yield a problem that is in some important sense equivalent: *Jill can paint a wall in 8 hours, and Jack can paint the same wall in 5 hours. How long would it take Jill and Jack to paint the wall if they worked together?* An equivalence class of problems is generated in this way by substituting variables, restricted to taking on positive real values, for the specific values used in the initial statement of the problem: *Jill can paint a wall in x hours, and Jack can paint the same wall in y hours. How long would it take Jill and Jack to paint the wall if they worked together?* Conversely, this generically stated problem can lead to a set of specific problems by substituting specific values for the numerically valued variables, x and y.

Similar equivalence transformations and associated forms of abstraction are defined for the other variable types mentioned. Furthermore, psl permits the specification of arbitrary type names, each of which implicitly defines an equivalence class.

Equivalence Under Substitutions of Predicates

In the preceding problems, the action *paint a wall* could be replaced by any other *same type* predicate, such as *mow a lawn,* without affecting the substance of the problem, as long as the semantic import of the predicate is taken to be the same (thus, perhaps, *make soup* would be a different predicate type and be deemed an inappropriate substitution).

An equivalence class of problems can be generated in this way by substituting predicate variables P, Q, etc. for the specific actions mentioned in the problem, restricting substitutions to the specified type of individual predicate. Conversely, the generically stated problem can lead to a set of specific problems by substituting specific actions for the typed predicate-valued variables, P and Q.

Equivalence Under Symmetry-preserving Permutations of Variables

Given a problem statement containing variables, such as the examples cited earlier, one can rename each variable without changing the essence of the problem as long as type restrictions are honored. That is, w can be replaced by x and y by z, provided all instances of w and y are so changed, that x and z are not used elsewhere, that w and x are of the same type, and that y and z are of the same type.

A more interesting set of renamings involves *permutations* of variable names. Consider the psl statement $p(\{Jack, Jill\}) = 3$. One reading of this is taken to be "the set of people with personal names Jack and Jill can accomplish a task in 3 time units." Assume that it is a property of the predicate p that the variables in its first set argument can be freely permuted, so that $p(\{Jill, Jack\}) = 3$ is equivalent to the first statement. Considering in addition the symmetry of the logical connective &, we then have the equivalence of the following four statements:

$$p(\{ \text{Jack,Jill} \}) = 3 \,\&\, p(\{ \text{Larry,Moe} \}) = 12$$
$$p(\{ \text{Jack,Jill} \}) = 3 \,\&\, p(\{ \text{Moe,Larry} \}) = 12$$
$$p(\{ \text{Jill,Jack} \}) = 3 \,\&\, p(\{ \text{Larry,Moe} \}) = 12$$
$$p(\{ \text{Jill,Jack} \}) = 3 \,\&\, p(\{ \text{Moe,Larry} \}) = 12.$$

These facts can be summarized by a set of equivalence-preserving permutations. There are five such permutations in addition to the identity permutation:

(Larry Moe)(Jack)(Jill)(3)(12)

(Larry)(Moe)(Jack Jill)(3)(12)

(Larry Moe)(Jack Jill)(3)(12)

(Larry Jack)(Moe Jill)(3 12)

(Larry Jill)(Moe Jack)(3 12).

The first four are sufficient for generation of the equivalence class of statements.

While these examples are obvious when made explicit in this simple case, in fact, they are not obvious to all aspiring problem solvers, and failure to recognize these equivalences can greatly hamper search of a problem graph. In more complex situations, when the number of variables increases and the number of equivalent statements multiplies, the set of equivalence-preserving transformations becomes difficult to enumerate.

This type of symmetry is particularly common in pictures. Gelernter (1959) constructed an algorithm for generating all "syntactically symmetric" statements (as he called them) by constructing the appropriate sets of permutations of variable names. His work was done in the context of a geometry theorem-proving machine, where equivalences under variable permutations reflect spatial symmetries. For example (see Figure 14.1), a theorem

If line AB is parallel to line CD, line AD is parallel to line BC, points A, C, and E are colinear, and points B, E, and D are colinear,

Then, the distance between A and E equals the distance between E and C, and the distance between B and E equals the distance between E and D.

The two clauses of the conclusion are symmetric in the sense that any proof of one of them can be converted to a proof of the other simply by an appropriate substitution of variable names. Gelernter constructed an algorithm to compute all permutations (substitutions) of the appropriate sort, based on given properties of the predicates *parallel, colinear,* and so on. Clause 10 of our definition of psl permits the specification of such symmetry-defining properties of predicates.

The Gelernter algorithm can be applied in other problem situations. Doig (1965) provides several examples. The algorithm, being an exhaustive generator of this kind of equivalence, can be used as a normative model against which to measure human novelty-detection skills in a variety of problem-solving tasks.

Topological Equivalence

In many types of problems, diagrams are used in presenting the problem and the solution attempts. Often, the scale and other geometric features of the problem are irrelevant to its essence, although particular values must be selected for presentation. For example, electrically equivalent circuit diagrams result from topology-preserving transformations that change scale or rearrange components on the page while preserving connectivity. In such cases, the connectivity properties of the diagram, being representable as first-order predicates (e.g., *connected to*), can be stated in a psl of the required form. For example, the circuit of Figure 14.2 can be described as follows, using the labels given in the figure and taking w(x,y) to denote that x and y are connected by a conductor:

w(1,2)

w(3,4)

w(3,7)

w(5,6)

w(5,8)

B(battery) = 6 volts

L(coil) = 100 microhenries

R(resistor) = 1000 ohms

C(capacitor) = 100 microfarads.

This description can be used to generate a variety of equivalent circuit diagrams, of which our description of Figure 14.2 is one example.

Other Problem Domains

Another class of problems in the scope of these analytic methods comprises simple puzzles in which visual arrangements are to be manipulated—for example, match stick problems, coin problems, and similar puzzles (Gardner 1982). Consider the nine dots problem. A three by three array of dots is presented with the instructions that a continuous line composed of four straight segments is to be drawn through all nine dots. If the dots are numbered as in Figure 14.3, solution attempts can be represented as sequences of dot names. For example, the attempt depicted in Figure 14.3 is < 1 2 3 6 9 8 7 >, and one correct solution is < 9 5 1 2 3 6 8 7 4 >. The symmetry of the problem can be captured by a set of permutations on dot names; for example, a 90° clockwise rotation of the board is achieved by the permutation (1 3 9 7)(2 6 8 4)(5). This permutation applied to the depicted solution attempt yields the equivalent attempt < 3 6 9 8 7 4 1 >.

Detection of Equivalence and Novelty

Once the transformations on a pattern that leave it unchanged are specified, it is not difficult to design an algorithm that applies them to facilitate recognition. Problems in a psl can be regarded as such patterns. If a linguistic, graphic, or mixed pattern denoting a problem appears to a problem solver P, the first thing P should do is to check if P has previously encountered problems in the same equivalence class. If the latter has been successfully solved, the solution may directly or indirectly be transferred to the problem at hand. If it has not been successfully solved, then previously tried attempts should be avoided.

Scientists are trained to apply systematically a number of standard transformations, such as rotations, translations, dilations, and so forth to geometric structures. Analogies are invariants under such transformations at higher levels.

Problem recognition is difficult when a small set of frequently useful transformations does not produce a familiar pattern, or when these transformations are not known or thought of. P would then learn them by applying randomly selected transformations or compositions of these and detecting recurrences. Novelty is detected either after a long and futile search for transformations or on encounter with a pattern constructed to or easily checked to be untransformable to a known pattern. Cantor's diagonalization method produces patterns constructed to differ from all previously encountered patterns.

Checking whether a problem presented to P is novel or is an equivalence class of previously encountered problems imposes both a memory and processing burden on P. If P is computer-based, the burden is easily met. If P is a living organism, acceptance of this burden may cost P the opportunity to do other, more survival-related activities. The ability to detect equivalences may thus not have evolved through natural selection. If P encounters very few problems very frequently, if P's survival depends on coping with them very quickly, and if problems whose equivalence can be determined only by applying many transformations are rarely encountered, then P cannot afford to waste time on routinely assigning to each problem its equivalence class.

P will evolve to adapt to this narrow problem environment without great ability to generalize, to transfer problem-solving skills, or to adapt to changing environments.

Summary and Conclusions

Our informal empirical investigations to date have confirmed our impressions that nonexpert problem solvers are poor at detecting equivalences in formal problems. Nonetheless, everyday problem solving of the sort called common sense appears to be relatively unhampered by the sort of wheelspinning, repetitive solution attempts that people are caught up in when solving technical problems. One reason for this, we conjecture, is that many of the formal equivalences that are opaque when stated algebraically or linguistically are readily apprehended when presented visually. Even the nine dots problem, though difficult in its usual presentation, is almost totally incomprehensible when presented in a linear, algebraic form of the sort illustrated in the preceding discussion. The 2-dimensional, visual world provides additional structure that is useful for clarifying equivalences (among other things, see Lindsay 1987).

Notes

1. Note that the number 9! = 362,880 denotes the number of paths through the problem graph conceived as a tree. This conception assumes that some board configurations will appear at many nodes in the tree, corresponding to different sequences of moves that lead to the same board configuration. A board configuration can be coded as a nine-digit ternary number (O, X, and blank). Three raised to the ninth power is 19,683, so this is the total number of possible distinct board configurations. If the graph is unified by representing each possible configuration only once, it ceases to be a tree and becomes a graph with 19,683 nodes. The number of possible paths through this graph, starting at the initial position and ending at one of the $2^9 = 512$ completely filled configurations, is still 362,880, corresponding to the 9! possible legal sequences of moves defined by the rules. This analysis ignores the fact that continuations following a win by either side are of no further interest and could be excluded from the graph (or tree). Our analysis of the game when symmetries are considered also ignores this additional simplification by ignoring post-win continuations, so the figures from the two analyses are comparable in this regard.

2. The enumeration of the symmetry-reduced graph is straightforward but tedious, and would serve no purpose in this exposition. As a simplification, it is easy to enumerate the symmetry-reduced tree to two levels. Ignoring the symmetry between players, the initial node leads to 3 distinct moves; two of

these lead to 5 distinct moves, and one leads to only 2 distinct replies, each of which, in turn, leads to 4 replies. To this depth, we thus have 10 nodes, each with 6 blank squares and 8 nodes, each with 5 blank squares. Ignoring further reductions by symmetry gives us an upper bound on the number of games in the game tree: $(10 \times 6!) + (8 \times 5!) = 8,160$.

3. Equivalence at a higher level can lead to even further reductions and simplifications. The three equivalence classes of configurations with three Xs in a horizontal row form a new equivalence class. So do the three vertical patterns. The invariants under what seem to be logically—as well as psycho-logically—simple transformations have been studied in the simplest possible problem (Kochen and Galanter 1958). The problem is to infer the next bit in a periodic sequence of bits, such as 011001100110 . . . or 00110011. . . . The essential repeating pattern, 0110 or 1001, can be regarded as 01 followed by its mirror image; alternatively, the pattern could be encoded in 12 (01 = 1, 10 = 2), and further patterns could be detected. The theory of patterns (Grenander 1978) could be brought to bear on such analyses.

Bibliography

Doig, L.J., 1965: Theory and use of the method of syntactic symmetries, Master's thesis, The University of Texas, Austin, Tex.

Gardner, M., 1982: *Aha! Gotcha: Paradoxes to Puzzle and Delight*, W.H. Freeman, New York.

Gelernter, H.L., 1959: Note on syntactic symmetry and the manipulation of formal systems by machine, *Information and Control* 2:80–89.

Grenander, U., 1978: *Pattern Analysis*, Springer-Verlag, New York.

Knuth, D.E., and R.W. Moore, 1975: An analysis of alpha-beta pruning, *Artificial Intelligence* 6:293–326.

Kochen, M., and E.H. Galanter, 1958: The acquisition and utilization of information in problem solving and thinking, *Information and Control* 1:267–288.

Lindsay, R.K., B.G. Buchanan, E.A. Feigenbaum, and J. Lederberg, 1980: *Applications of Artificial Intelligence for Organic Chemistry—The DENDRAL Project*, McGraw-Hill, New York.

Lindsay, R.K., 1987: Images and inference (unpublished manuscript).

Newell, A., J.C. Shaw, and H.A. Simon, 1958: Chess-playing programs and the problem of complexity, *IBM Journal of Research and Development* 2:320–355.

Newell, A., and H.A. Simon, 1972: *Human Problem Solving*, Prentice-Hall, Englewood Cliffs, N.J.

Nilsson, N., 1971, *Problem-solving Methods in Artificial Intelligence*, McGraw-Hill, New York.

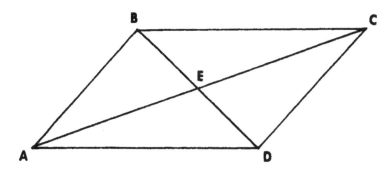

FIGURE 14.1 Diagram to illustrate Gelernter's Algorithm

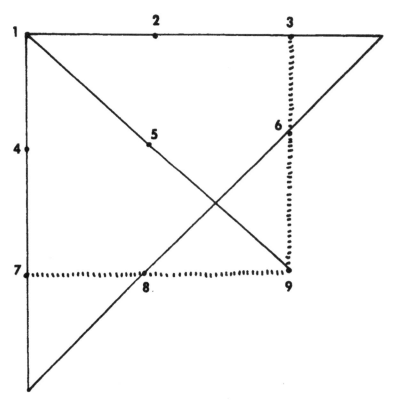

FIGURE 14.2 Circuit

240

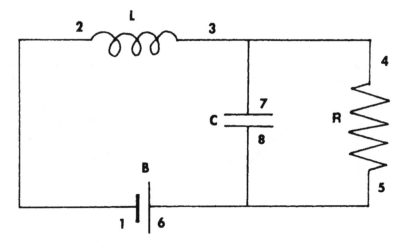

FIGURE 14.3   The Nine Dots Problem

15. Order and Disorder in Knowledge Structures

Introduction

The question of how brain becomes mind is central to cognitive science.[1] Cognition refers to the process of knowing and the processing of knowledge. Cognitive processes are involved in such human intellectual activities as "understanding," "thinking," "knowing," "discovering," "being aware of," "reasoning," "judging," "perceiving," and "imagining." All these are functions of mind. They are involved when we recognize a face, understand an article, or experience a discovery. Explaining how such higher mental functions can be and are performed is one of the outstanding challenges to science at one of its current frontiers.

Explanation can be sought in several ways. We can ask how these functions are implemented in the nervous systems and brains of living organisms. We can also ask how these functions could be implemented by a machine, unless that is proved to be impossible. We can inquire into the theoretical possibility—i.e., try to prove or disprove the mathematical existence—of an automaton programmed to observe its own actions, classify objects or events of experience into conceptual categories, record and recall such encoded experiences, and anticipate previously unencountered experiences. We also ask if such an automaton could recognize similarities, identify functional equivalence, apply known procedures to new situations, and detect regularities in a sequence of instances. These three modes of search for explanation should be integrated when the behavior of an automaton is claimed to simulate the behavior of some living embodiment of mind, or when the behavior of such a machine is compared with the behavior of a living system coping with the same sample of tasks.

The challenge to a designer who would embody cognition in a machine is not nearly as great when there are: (1) a few well-defined tasks; (2) clear criteria of what constitutes an adequate response that can be produced by searching a well-structured knowledge base appropriate to the tasks; (3) a good inference procedure; and (4) reliable means of dealing with inputs and outputs that are English-like or subsets of a natural language. But the challenge of organizing knowledge so that it can be brought to bear, as well as added to, is enormous when the number of possible tasks it has to cope with is very large and when they are ill-defined.

The brain does this last task effectively. It has been this author's thesis that the underlying principles are simple. This is both a starting assumption to guide the inquiry and a working hypothesis for which evidence is to be sought. If this is correct, it inspires a sense of wonder about how the organization of knowledge is fundamentally linked to the discovery process, to uncovering new relationships. There is also an impasse about how vast, complex bodies of non-expert world knowledge could be represented in a machine, despite the successes of knowledge engineering (Groner et al. 1982; Elithorn and Banerji 1982; Mulsant and Servan-Schreiber 1983) in building expert systems (Davis 1982; Barr and Feigenbaum 1982; Shortliffe and Duda 1983). At least this author is motivated by these challenges to search for fresh ideas.

A fundamental controversy has divided researchers in this field. The vast majority have abandoned the search for simple, powerful, and elegant general laws. They believe the attempt would prove fruitless. They believe knowledge is processed by a great variety of diverse procedures and heuristics, rather than according to a grand design and a few fundamental principles.

A common belief is that a compilation of thousands of hindsights and rules of thumb in the form of "if situation X prevails, do Y" is the basis of intelligence and thought. Even if such hindsight is the result of inductive algorithms operating on vast bodies of accumulated knowledge and prior heuristics, and even if it produces world champions of various games, it does not meet the criteria for the intellectually satisfying explanations that we seek. Heuristics or hindsight can be arbitrary and trivial, and mere compilation or accumulation of them does not constitute a theory.

A small group of researchers have persistently continued the search for a few, powerful basic principles in the direction of adaptive learning involving the continual reorganization of knowledge. (Kochen 1960a; Doran 1968; Gold 1971; Andreae 1974–77). In this chapter, one line of inquiry based on this approach is presented (Kochen 1971, 1974).

Background

This line of inquiry began with the analysis of an algorithm for concept learning based on a paradigm introduced by Bruner, Goodnow, and Austin (1956) (Kochen 1960b). Input to the computer program implementing the algorithm consisted of a random sample of n-bit words. As an example, consider three 3-bit words, 010, 011, and 101. Interpret them as three objects, each described by the same three two-valued attributes, such as size (0 = small, 1 = large), shape (0 = circular, 1 = square), and color (0 = blue, 1 = red). For each object, indicate whether or not it is an exemplar of a "conjunctive concept," such as "all small, red objects," for example, with YES, NO, YES for the preceding three objects. Output of the program consisted of hypotheses of the form X10. Interpret this as a "guess" that the size is irrelevant (X), shape is square (1), and color is blue (0), together with a weight reflecting the strength of evidence for that hypothesis. This was extended to "disjunctive hypotheses" or general Boolean combinations such as "X10 or 0X1," and used to compare the efficiency and effectiveness of inductive and deductive procedures, the effects of complexity in the task and of limited memory. This paradigm is still sufficiently general to subsume current research in "playful behavior," where one or more of the n bits encodes an action the program can perform on the object characterized by the remaining bits, and the YES/NO bit denotes the object's response to that action (e.g., Does it move?). It is, of course, possible to encode arbitrarily many—and many-valued—attributes, actions, consequences, with the use of more than one bit. The limitations of this paradigm were recognized quite soon, and led to a successor that was essentially a relational data model called AMNIPS (Kochen 1962), which contains many features found in some of the currently popular systems such as K-LONE (Brachman et al. 1978), DAPLEX (Shipman 1981), and SYSTEM R (Astrahan et al. 1976). Users of AMNIPS could specify whatever predicates, in place of two-valued attributes, they wished to use in describing the instances of a concept. For example, "a square" could be described by the predicate "has a number of edges that is 4." In contemporary list-processing structures (Winston and Horn 1981), such a structure is represented as a nested association list with the node "large square" an instance of a "slot name," and the "four" that fills the second slot is a "value." In AMNIPS, the "value" was a node in its own right. In the representation to be described in this chapter, nodes, predicates, or links (slot names in "frame-theory" parlance) are all to be regarded as nodes (frame names) at some other level. That is, "slots" in a structure could be filled by other structures in this 1962 system. The use of LISP, by virtue of its internal structure and associated procedures, makes programming and execution far more efficient than was the case for this 1962 system. Both deductive and inductive procedures were implemented to a limited extent in that early system.

In 1973, it was possible to prove theorems (Hantler and Kochen 1973) about the speed with which a distribution of weights over hypotheses about a simulated "learner" and its environment consisted of objects that moved according to some "laws of motion" that could be arbitrarily chosen by the investigator. The simulated learner received the equivalent of observations about the positions and motions of these objects, or about the outcome of experiments by the learner. One of the objects in the simulated environment was intended by the investigator to represent the physical embodiment of the learner itself. That is, the simulated learner produced actions that controlled the motion of that special object. When the object denoting the learner was in some relation (arbitrarily specified by the investigator) with the other objects of the environment, the simulated learner was rewarded. Later, a reinforcement gradient related to proximity of the present state to a possible goal-state was introduced. In the first set of simulations, the learner was supplied, as input from the investigator, with a set of equally weighted hypotheses about the laws of motion of the environmental objects and about the values of environmental states to the learner. The program simulating the learner had to select hypotheses, use them to select a goal and actions, and use feedback about the consequences of each action to modify the distribution of weights associated with the hypotheses.

The next step along this line of inquiry was to enable the simulated learner to form its own hypotheses from a given generative grammar (production rules forming production rules). By 1975, it was possible for the simulated learner to form hypotheses about the regularities in sequences such as the following: "every third beginning with the first are the odd numbers 1,3,5, . . . ; every third beginning with the second are the squares, 1,4,9,16, . . . ; every third beginning with the third are the even numbers 2,4,6, . . . " (Kochen 1975). This came close to the essence of what other induction programs are now doing on a more ambitious scale (Gardner 1983). To be sure, such simulations still lack the realities that faced discoverers at the time of discovery: how to select and purify the data from the wealth of natural distractions to which to pay attention.[2]

More important, however, was the question of the genesis of predicates and concepts. By 1978 (Kochen and Stark 1978; Kugel 1977; Blum and Blum 1975), the focus had shifted to the possibility of a general induction algorithm, including the formation of concepts appropriate to an arbitrary environment. Rather strong assumptions or conditions about the compatibility of the environment and the learner had to hold for such an algorithm to exist. A completely general induction algorithm has been proved not to exist, to be impossible. The growth and organization of human knowledge throughout history does, however, attest to the existence of some process—perhaps not a specifiable algorithm—for generating concepts.

The limitations of this approach have now also been recognized, and have led to a renewed[3] emphasis on structures for representing what is known (Kochen 1982) in ways that may make possible the comprehension and discovery of novel structures in restricted domains. In what follows, we present our current conceptualization of these issues, first by examining the importance of representation, second, by defining structure, third, by looking at the role of ordered structures in discovery processes, and finally, by discussing the dynamics by which ordered structures are formed and used. This culminates in the proposition that knowledge increases and organizes itself by "autopoietic" principles stimulated by the metaphors of thermodynamics.

Representations

Whether a few powerful basic laws help the scientific community explain cognition depends critically on how this community represents or conceptualizes the issues. One reason for the existence of powerful laws in physics is that ideas, such as motion, were represented by fruitful concepts, such as momentum. One of the key ideas to be conceptualized here is the notion of representation itself. The importance of selecting an appropriate representation has been well illustrated in the study of problem solving, particularly by computer. The ability to select or shift to an appropriate system of representation is critical for bringing elegant and powerful analysis methods to bear. In solving mathematical problems, it is common to transform or reformulate a problem so that a known method can be applied, or so that understanding is greater or so that the solution becomes obvious.

Formal definitions of "representation" and "system of representation" have been proposed (Kochen 1982), with hypotheses as basic ingredients. In this chapter, the fundamental unit of representation is taken to be a "structure," what Lenat (1982) and Thagard (1983) call a conceptual frame, what Margenau has long ago called a "construct," and what shall here be called a knowledge structure.

Certain properties of a structure should remain unchanged, even if there is translation from one representation to another. That includes translations from representations at the level used for public discourse to representations at the level used for private contemplation, and even to representations at the level of neural events that may be correlated with knowledge structures. Possibly there is a canonical representation at each level. Desiderata for such a representation include appropriateness to the environment, as well as to the procedures available for coping with it. Desiderata for a system of representation include ability to shift representations readily in the direction of appropriateness.

Consider some examples of representations. If a person is asked to state all he can think of about his knowledge of a straight line, the following list of responses results:

1. shortest distance between two points,
2. motion from one point to another,
3. a set of points,
4. indicates a direction,
5. an edge,
6. a divider (between two regions),
7. could be parallel to another line,
8. could intersect another line,
9. could be skew to another line in space, and
10. slants or slopes.

A more complete list is contained in the appendix. Many respondents are likely to select (1), partly because they have at one time learned this as a definition or phrase in English or some other natural language whose terms conventionally indicate some intention for a speaker or writer. All are phrases in a language for communication among persons about cultural artifacts, such as the string of a plumb-bob. By contrast, the mental images reflected by such a linguistic representation or those which are unique to one respondent are often private, nonconsensual, internal symbols—for example, the mental construct of a line. Lower still than these two levels is a neurodynamic level with biochemical and neurophysiological events underlying the mental states. Examples are the columns of cortical cells that respond to lines at different orientations.

Students familiar with analytic geometry, when asked to express what they know about straight lines, might say a line is the set of all points (x,y) such that $ax + by = 1$ for some pair of numbers a,b. They might also recall or derive that it is b = the set of all polar coordinates (r,u) such that $1/r = (a \cos u + b \sin u)$. These are two different representations by the same person of the same object at the same level (2) of representation. To such a student, the meaning of "straight line" is what is invariant under transformations among all his represenations, and this meaning deepens as he becomes acquainted with more representations and the recognition of transformations that render them equivalent.

A Cartesian coordinate system is a *system of representation.* It provides for the formation of many *representations,* such as ax + by = 1 for a line, $x^2 + y^2 = r^2$ for a circle, etc. The polar coordinate system is another system of representation. It, too, provides for the formation of numerous representations. The transformation from the polar coordinate system to rectangular coordinates is give by x = r cos u, y = r sin u, and this can be applied to any representation to check equivalence.

Expressions that are simple and in which certain patterns are detected relatively easily in one system of representation may be very complex and resistant to discovery in another system of representation. For example, the representation of a line through the origin in Cartesian coordinates is simple, and reveals readily that the point (kx, ky) lies on the same line as (x,y) for any number k. In polar coordinates, the expression is cumbersome and does not reveal this so readily. On the other hand, the equation of a circle, r = constant, or spiral, r = ku, is very simple and revealing.

Numbers written in different bases provide another example of different representation systems. The representation of the decimal number 9 in the binary system is 1001 ($1 \times 2^3 + 0 \times 2^2 + 0 \times 2^1 + 1 \times 2^0 = 8 + 0 + 0 + 1 = 9$). Suppose we had five bottles of 1,000 pills, some of which were filled with pills which were 5 milligrams too heavy. In one bottle, all pills are of equal weight. What is the minimum number of weighings needed to identify each defective bottle? It can be done in just one weighing. Take one pill from the first bottle, 2 from the second, 4 from the third, 8 from the fourth, 16 from the fifth. Weigh all these pills that were removed together. Suppose all these were 45 milligrams more than they should be. The *number* of such faulty pills must be 45/5 = 9, which is 1,001 in binary representation. One pill came from the first bottle (the 1 in position 1 of 1,001), and eight came from the fourth. Thus, the defective bottles have been identified to be #1 and #4 in just one weighing, thanks to the binary representation.

A favorite test vehicle for demonstrating how representation affects the search for solutions to puzzles is the Tower of Hanoi puzzle. The task is to transform the situation shown in Figure 15.1 to Figure 15.2 by moving one disc at a time from one peg to another, without ever placing a larger disc on top of a smaller one. Without a good representation, this problem, like the Rubik cube, can be very challenging when there are many more than three discs. Figure 15.3 represents all possible moves if there were only one disc. Each corner of the triangle denotes a possible location of the disc, and an edge denotes moving the disc from one peg to another. Thus, if the lone disc was initially at A and the goal was to move it to C, a single move from A to C would suffice. For 2 discs, consider Figure 15.4. Here, vertex A means that both discs are at A (the smaller on top), and it is denoted in parentheses (location of smallest disc, location of largest disc). This gives all possible moves from each possible starting point to each possible end point. Figure 15.5 gives it for the 3 discs. The starting point as given in Figure 15.1 is circled and denoted by (B,A,A) to be interpreted as: (the smallest disc on peg B, the medium one at A, largest at A) and the goal is point (C,C,C) corresponding to Figure 15.2, with the optimal solution shown by the darkened line in Figure 15.5 (7 moves).

It is easy to see how the representation changes as one more disc is added. It is also easy to see how to find the best solution for any problem that can be posed for this puzzle.

Programming computers to solve puzzles such as the Tower of Hanoi has become a test vehicle in exploring the properties of various algorithms believed to characterize some aspects of intelligence. It is also common to represent the knowledge needed to "reason" about actions in "frames" that are inputs to LISP programs. Programmers have, however, tended to associate to each algorithm a particular data structure—which has been loosely identified with a representation—and some have regarded the search for an effective representation that is independent of the exact problem to be solved as an "ill-posed activity." Such activities are essential in discovery process and in ordering knowledge structures.

In AM (Lenat 1976; Davis and Lenat 1982), a program that represents mathematical concepts and is governed by heuristic rules to discover interesting mathematical conjectures, each concept is represented by a fixed set of facets. AM uses 25 facets, not all of which apply to every concept. Examples of facets of a concept are: concept name, concepts having less restrictive definitions than C, concepts having more restrictive definitions than C, examples of C, concepts of which C is an example, definitions of C, potential theorems involving C, analogous concepts, etc. Syntactically, a facet-value pair is part of the "property list of a LISP atom" naming the concept, one feature-attribute of which it describes. (The author and Paul Resnick have worked on a program called PM to supplant AM and to fit into the line of inquiry described here (Kochen and Resnick 1987).)

Such nested association lists do not provide for the symmetries that we regard as essential, particularly the dual role of every structure by which it can both be part of other structures and contain other structures as part of itself. For this purpose, it will be convenient to regard a structure at any level as a node or vertex. At a lower level, the same structure may be a topological simplex (Atkin 1974). For example, a large, red square can be represented topologically by a triangle (n = 3 attributes), with an edge linking two nodes if they are shared by the entire concept or object as in Figure 15.6a. If we considered a large, red, wooden square, with n = 4, the simplex to represent it would be a tetrahedron shown in Figure 15.6b.

Structures

In this section, we recursively define knowledge structures, or structures, for short. We start with a specified system as illustrated before, a subset of English, a programming language such as LISP, or topological complexes. Let us use a subset of English for purposes of this presentation, although we later switch to the language of graph theory and topology. We start with representations of primitive structures, such as the terms "line segment," "_____ intersects _____," "_____ is called _____," "_____ is an example of _____," "_____ is exemplified by _____," "_____ is specific to _____," "_____ is generic to _____." We then specify syntactic rules for forming structures from structures. Thus, certain structures replacing the blanks of certain other structures are structures again. For example, "line segment is exemplified by line segment a" is a structure, as is "line segment a intersects line segment b." The latter structure can be inserted in the first blank in "_____ at _____." The second blank could be filled in by "point P," "some point," "some set," "point x for some."

For each of these structures in a linguistic representation, there is a corresponding graphic and other representation if the other systems of representation are sufficiently rich. Several systems of representation co-occur,[4] indicating a line segment both linguistically and graphically, for example. All that a person thinks of when his attention is focussed on "straight" lines is thus represented by a structure that may be more completely revealed by the length of time he concentrates. To concentrate means to keep attention focussed on a circumscribed set of structures (e.g., "line segment a") and reconstruct the structures of which it is part. *To think is to control attention,* directing it to access specified structures according to a program, or perhaps in a seemingly random way when no program has as yet emerged.

If we take literally the term "feature" as it is loosely used both in artificial intelligence and neurobiology, then we may, with Edelman, regard the brain to contain neuronal groups of 50 to 10,000 neurons that act as detectors of a primary repertoire of features. Thus, there are special neural structures that detect the presence of straight lines in an animal's visual field, a column for lines of each slope class, corresponding to slopes about 15° apart. There are also recognizers of recognizers—a secondary repertoire—that, following principles of natural selection, generate associations among the recognizers. It is possible that the nodes or structures we have referred to are embodied as particular neuronal groups (see Chapter 4, this volume).

Because the complexity of such structures seems to be beyond our comprehension, we seek a simple representation of knowledge structure that captures what is essential about it for the purpose of explaining how knowledge is added. Because the number of conceptual structures or nodes is so large, it is tempting, for our purpose, to assume them to be interchangeable or identical, so that they can be aggregated and represented statistically, since statistical representation is simpler. But the nodes are unique almost to the extent that no two are identical or interchangeable. Yet, they are organized into patterns whose symmetries introduce order that provides the simplification sought for.

Consider a simple analogy: imagine a line of $n = 10^{23}$ particles whose spin can be 1 (clockwise) or -1 (counterclockwise). A statistical description might assert that the probability of any k of the n particles having spin 1 is (n), which is maximum when $k = n/2$. If the position of the particles mattered, another simple description is: "Every second particle has spin 1." This is highly ordered with many symmetries.

Why is simplification important? If the number of natural chemical elements were 100,000 rather than about 100, and if there were no periodic table, and if the number of types of force were 40 rather than 4 or 3 or 2, if the number of basic physical variables were in the 1000's rather than mass energy, space, time, charge, etc., then expressing the laws of motion might require several pages rather than "force = rate of change of momentum." Adding to or using knowledge then would, at best, be very slow, and unreliable even with the help of powerful computers.[5]

The central problem in the organizing and study of knowledge is to find representations that reveal order-producing symmetries that make possible the process of adding to. The simplest symmetries are illustrated by such conceptual structures as "_____ intersects _____." If we denote such a relation by R(x,y) then $R(x,y) = R(y,x)$. This asserts invariance of R(x,y) under the transformation of permuting (x,y). A more subtle kind of symmetry is one that allows one node to be substituted for another, as Mark Twain for Samuel Clemens in the sense that both occupy nearly the same place in a larger structure.

To understand easily a new conceptual structure is to connect it into the existing structure without changing the order of the structure greatly. A new conceptual structure is difficult to understand if it cannot be connected into the existing structure without making major changes in the latter. When such a reordering occurs, some symmetries may be broken, and important new structures may emerge.

Discovery Processes

A body of knowledge is enriched by organization and by being added to. Organization results in increasing the order of structures. The number of structures increases when new structures are uncovered or constructed.

What does it mean for a conceptual structure to be enriched, and how can this occur? A conceptual structure is impoverished if it is linked with very few other concepts or with concepts that are themselves impoverished. If an isolated node could exist, that would be an extreme. Yet there may be a node linked to just a few other nodes that are themselves weak, but the linkage thus created gives rise to a powerful new concept. Then, there are concepts linked to many others. But that alone does not make it enriched or indicative of meaningful knowledge. If the resulting structure is such that there are many patterns, actual and potential, then it is enriched. It may be rich in implications or in abductions (from one structure, infer that another structure is probably, possibly or plausibly true). For example, if "____ is called ___" links not only x and y but y and z, for any x, y, and z, and the link is stated to be transitive, then its application to x and z is implied. If it is applied, it becomes part of the structure, and an actual new triangular pattern enriches it; if it is not implied, the pattern is potential, but the structure is nonetheless richer.

The meaning of a conceptual structure to a thinker during a given time period is the part of the structure to which he attends. Insofar as parts of the internal structures of various thinkers correspond to one another, those aspects of meaning are shared and may be conventionalized so that the names of certain conceptual structures have standard meanings (core denotations and core connotations).

As a conceptual structure becomes enriched or becomes meaningful knowledge, it can generate uncertainty in the thinker. The uncertainty can refer to the possibility of a new structural aspect.

An impoverished or weak structure offers few or no possibilities for "new" structures or emergent structural properties. A rich structure has many such possibilities. Some of these possibilities are more readily accessible than others. Uncertainty is greatest if there are many possible "new" structures, each of which is equally inaccessible.

New structures are those previously unattended to but in actual existence or constructed from existing structures by combination, composition, correspondence, and other operators. If the operators were all specified, then every new structure has potential existence. But the system of operators and operands is never completely specified, with new elements emerging due to environmental stimulation.

The action of reducing such uncertainty is informing (or being informed). This is best understood as filling open slots (de Mey 1982). Information may flow from an agent external to the thinker to him or from his own thought process in response to his question.

A question reflects the focus of attention on some anomaly, some departure from structures that are common, readily accessible. Uncertainty may induce question formation. Insofar as structures have logical or semantic properties, the anomalies may take the form of contradictions, gaps, ambiguities. Having raised a question is likely to be a precursor for discovery. Readiness to notice novelty is another characteristic of a mind prepared to make discoveries.

An example of a structure that evokes uncertainty is a dimly perceived human figure emerging from the dark. Is it a man or a woman? Is it a threat? Do I know the person? These may be some of the questions that could arise. As the figure appears in the light, information that reduces uncertainty is received. Without the appearance of the figure, there would have been no uncertainty in the first place. Note that, as in the mathematical theory of communication, information is viewed as the reduction of uncertainty and is devoid of meaning when the conceptual structure is impoverished, but not so much that no a priori uncertainty exists. If the structure is enriched, the possible questions become interconnected with one another and other conceptual structures as well, and a distinction between profound and trivial questions begins to emerge. Profound questions tend to lead to important discoveries.

Indeed, the recognition of and commitment to important questions is what distinguishes understanding from knowing. When a thinker not only knows answers to meaningful questions, but recognizes what he does not know, as well as its importance, we regard this as an operational definition of understanding. Under normal conditions, he will commit himself to acquiring knowledge that he considers to be important. Commitment is, of course, necessary, but not sufficient. The resources that are committed, including attention, must also be sufficiently powerful, concentrated, and well-managed. In discovery, there is an alternation between concentrated and relaxed or playful thinking.

The form serves to increasingly prepare a mind to notice novelty, to make discoveries.

When thinkers ask themselves a question, then answer it, they acquire knowledge by recall, inference, or discovery. In the broad sense that knowledge acquisition uncovers potential or hidden (covered) patterns already present, discovery includes both recall and inference. That is, all structures to be attended to or constructed are already explicit in the system that specifies the construction procedures except for the generation of new operators, the procedure for which is not prespecified. In the narrower sense that it constructs new patterns (that are, nonetheless, potential and feasible within a given structure), discovery—or invention—is distinct from recall and inference.

The process of discovery consists of procedures that operate on conceptual structures, as well as on the world external to the thinker. Some procedures, such as observation, search for and notice previously unencountered patterns in the external world. Thus, an artist can "notice" patterns in clouds or trees that the non-artist does not notice. Other procedures, such as pure thinking, search for and notice previously unencountered patterns in the thinker's internal knowledge structure.

These procedures are organized into serial, as well as parallel, schedules according to higher order procedures. As mentioned previously, thinking is taken to mean the control of attention to nodes in the knowledge structure of a thinker. Attention is a procedure for following trails in the knowledge structure, possibly several simultaneously, starting at a given node. A high-order procedure may confine attention to remain within a certain number of removes from the starting node for a given time period. Thinking is deliberate or planned if a higher order procedure, using at least a local and imperfect map of the knowledge structure, instructs attention to shift only over selected trails, guided by an objective. Thinking is concentrated if it is deliberate, and focussed on one objective to the exclusion of all others. It is playful if attention is unconstrained, not even adhering to a requirement to follow a trail continuously. Exploratory thinking is more constrained than playful thinking, but less so than deliberate thinking.

The statement that chance favors the prepared mind is attributed to Pasteur. To explicate it in this context, we regard a mind as a representation or a map of a knowledge structure—i.e., of a representation. To be prepared (for noticing a previously undetected pattern that is also felt to be important) means that the representation of representation can shift among the latter. If no pattern is noticed in one representation, another representation is tried. But not just any other one. The representation or map of representations guides the choice, both according to "heuristics" as well as selection laws.

The use of programmer-supplied heuristics has been common in artificial intelligence since its start. They are rules of the form "if situation ＿＿ prevails, do action ＿＿＿," and they represent the compiled hindsight of the programmer and experts whose brains s/he "picks." But hindsight-based production rules cannot fully explain discoveries in the sense of noticing previously unnoticed patterns, because the system has to be told what patterns to look for. Moreover, a thinker is more likely to use rules of the form "if the situation is ＿＿ and action ＿＿ is taken, then change ＿＿ is ＿＿ likely to take place, and that change is favorable to degree ＿＿." Of course, MYCIN does give weighted judgments of this kind. The use of facet names to specify the role of a slot name, such as color, could be extended to incorporate distributedness over a range of colors. Still, the central process of generalization—and perhaps of analogy—is more difficult to simulate with production rules than with frames. Production rules act largely independently of one another. That is why PM—a Plausible Mathemachine—has been programmed in the frame-based, object-oriented system, KEE.

At the physiological level, the noticing of hidden patterns is much more likely to occur by natural selection mechanisms. Thus, after a neuronal group's ability to respond to a "right triangle" has been amplified sufficiently for it to recognize many right triangles, the right triangle ABC embedded in Figure 15.7 could be noticed; so could the squares CDEA, EHGF, and ABIF, if a square-recognition group has "evolved." Programs in LISP, PASCAL, and PROLOG have been written by Blaivas and by West for representing and processing knowledge about these geometric concepts, with a view toward discovering the Pythagorean Theorem and/or any one of its 460 proofs. Finding the proof implicit in Figure 15.7, for example, involves discovering the embedded square ABIF and that it is the square with the hypotenuse of right triangle ABC on its side, and that the four right triangles surrounding ABIF that fill out the remainder of the area of the large square have the same areas as the two remaining rectangles in the large square. A greater challenge, of course, is to construct the large square given right triangle ABC in playful thinking with a view toward reproducing the discovery of this ancient insight. While these programs do not exhibit anything like discovery, the program PM written by Resnick in KEE comes closer in that it demonstrates the power of Polya's patterns of plausible inference and of localized searching in knowledge space.

Discovery need not occur at a conscious level only. We propose that a thinking (cognitive) process is at a conscious level if it uses a representation of its own representation, particularly representations of external inputs, and if it can register such input representations in both the conceptual structures and in the map of these. Note that we did not require that inputs be registered at *appropriate* conceptual structures. Inputs could be from internal sources too, however; concentrated and deliberate thinking is conscious thinking.

But discoveries can result from playful thinking, too. The results of such discoveries may be dormant for a long time in the sense that they are not registered in both the conceptual structures and maps to these. If it is registered in a conceptual structure only, it can then be found, but not by a deliberate, map-guided search. Nor can it be "thought of" as having been encountered. If it is not registered at all, it cannot be found. If it is registered on the map but not in the conceptual structures to which the map refers, it may be thought of, but not found.

Discoveries, like other patterns in a conceptual structure, can be imaginative or dull. Imagining, we propose, is the procedure for recalling past "experiences" (i.e., conceptual structures) *or* anticipating unencountered ones that do not present themselves as concrete inputs (e.g., as line segment a, but imagined as any line segment). It involves the processes of abstraction, generalization, and conceptual combination, particularly by construction operators, in an essential way. Thus, a person who knows the concept of circle can mentally construct a torus by sweeping a small circle at right angles all around the latter. He uses no particular small or large circle in this mental construction. Having formed a structure for this new concept, he would "recognize" a torus if he saw one as an external input for the first time by mechanisms suggested in the next section. But the discovery occurred when the torus was first imagined. A pattern is imaginative to the extent that it differs from concrete inputs that have been encountered. The issue of how genuinely novel structures can be constructed or noticed has been alluded to. Insofar as the system for specifying structures and how they are generated is completely and precisely specified, all structures are implicit. As in predicate logic, however, our ability to generate structures outruns our ability to anticipate the outcomes of that generating process. In the next section are some ideas for how this process of generating new structures and new operators could occur. This leads to the idea of knowledge space, consisting of niches for new concepts, with emphasis on how new niches can originate.

Epistodynamics

That two thinkers can communicate indicates common aspects of their knowledge structures. One level at which to explicate this is the neurodynamic one, involving the previously mentioned neuronal groups and principles of selection from enough pre-existing diversity. But by means of communication among individuals, communities of thinkers/discoverers enable all to add to their knowledge structure the discoveries of many of the others, and thus amplify the enrichment process. Consequently, the shared public knowledge structures may grow faster than the internal, private ones—perhaps even beyond the capacity of individuals to keep up.

To obtain some clarity in the relation between cultural and individual knowledge structures, their interactions, and their relative growth rates, it is useful to explicate the notion of a "knowledge space" (McGill 1976). This notion is intended to capture what all knowledge structures have in common, what is invariant under transformations between one representation and another. Its details are, of course, unknowable; minds represent only representations. The use of this notion in the abstract may, however, help shed light on the dynamics by which areas of knowledge merge, emerge, and die out. The representation of such public knowledge is the published record, and features of knowledge space can perhaps be estimated by publication counts, citation, and co-citation clusters.

A portion of such a public knowledge space is mapped into the corresponding knowledge space in an individual mind. The notion of knowledge space is thus general, and overlaps considerably with the set of all possible structures considered previously. A structure has been regarded as a discrete object. Each node can be replaced by another structure, without a limit to this process. Suppose that there were only one initial node, and it is replaced by two nodes as a first step in this process. The result is $1 + 2$ distinct nodes. After n steps, there are $1 + 2 + 2^2 + \ldots + 2^n = 2^{n-1}$ distinct nodes. As n approaches infinity, the result is a noncountably infinite set of nodes. We therefore regard a knowledge space as a smooth or continuous geometric object in which the discrete structures are points or regions.

The elements of a knowledge space are representations of knowledge structures of varying degrees of order. A knowledge structure with high, long-range order has many symmetries and low algorithmic complexity. The latter is measured by the smallest number of bits needed to describe the structure to a universal Turing machine (Chaitin 1977). More specifically, let s be a string of symbols, as in an English sentence, that expresses the solution to a specified problem or task, T. The task could be more specific and well-defined, such as ascertaining the truth of a mathematical conjecture, or general and ill structured, such as characterizing a specialty topic in science. Let L denote the length of the shortest s, given the most compact description, c(P,T), of both the task T and the problem solver, P. If T and P are well understood, then c(P,T) encodes all the information needed to find the bits of s describing a solution. L is the conditional entropy or algorithmic complexity of a knowledge structure. The discoverer's level of preparedness for recognizing a solution can be estimated by the computational complexity of finding a solution description, for which Levin (Solomonoff 1964, 1982) has found an upper bound to be 2^n c where n = the number of bits in the description of s, given c(P,T) and c = the computational cost of testing s. Now consider a knowledge structure—such as a red, large square—as one of a set of points in a space on which some distance function is defined. The set of all these points may comprise a 2-dimensional surface in 3-dimensional space (e.g., Figure 15.8a). The space is itself a knowledge structure, one that encompasses all the others. The distances between the points may, however, change. It is useful to interpret the elevation of the surface as potential energy.

Each point on this surface corresponds to a potential node in the representation of a knowledge structure. When some primitive conceptual structures are formed, perhaps according to a genetically coded program, perhaps according to a selectionistic procedure by which neuronal groups become associated by concepts, imagine the surface to become continuously dented in the neighborhood of the appropriate point. The resulting wells can be interpreted as potential energy wells, and these may eventually be connected by canals formed by flows, to use a water analogy.

Concretely, consider the cube P presented as input to the visual system, as shown in the middle of the left part of Figure 15.9.

Imagine the darkness or intensity of the 12 line segments of this drawing to vary from 0 to 1, with lines a,b,c having darkness 1, lines d,e,f,g,h, and i having value 1/2, and lines j,k,l with a value close to 0 (barely visible). Suppose this structure corresponds to point P on the surface (see Figure 15.8a). Cut the surface vertically along a plane P, and examine the cross-section in the neighborhood of P (Figure 15.8b).

Now decrease the darkness x of lines a,b,c at the same time that the darkness, 1 − x, of lines j,k,l is increased. Let x also be the distance along the circular meridian of the torus, starting from P. When x = 1, the figure is now as shown, corresponding to point Q on the cross-section (Figure 15.8c). A figure in-between, such as the ambiguous

Necker cube for x = 1/2, is halfway between P and Q (see center of Figure 15.9 and point N in Figure 15.8c). Patterns at points to the left of N, if jiggled ever so little from equilibrium, are attracted to or fall into the well at P. Patterns do not, of course, move or fall. It is our attention to them that does.

The dynamic we propose is as follows. As attention to perceived or imagined structures shifts and "falls into a well," the well deepens and widens. In this way, the initially smooth surface becomes cratered (still smooth, but less so than before). Eventually, a saddlelike valley will form connecting P and Q.

If the cube is deformed by lengthening or shrinking segments, say h and k, by a factor of y, so that the length of h' is (1 + y) (where the length of h is 1, with −1 < y < 1), then figures P and Q are transformed as shown into top and bottom rows of Figure 15.9. While P can be transformed into Q or into P' or P'', and Q into P, Q', or Q'' simply by varying x and y, P' cannot be directly transformed into Q'. Since P' or P'' are encountered less frequently than P, a ridge from the points denoting P' and P'' sloping down into the well at P is likely to form, perhaps along a dimension perpendicular to the line joining P and Q. Thus, a canal network may form.

The representations corresponding to a knowledge structure in a mind share with this geometric representation only the relative location of the wells and canals, as well as the stability and bifurcation character of each point on them. A bifurcation point is one that changes its character from a stable equilibrium point to one that is unstable, generally splitting it into two new stationary points that are stable, while some parameter is changed continously. As an example, suppose that the potential energy associated with the dents and canals on the kind of surfaces (knowledge space) that we have discussed were describable by a quartic equation: $V(x) = x^4/4 + ux^2/2 + vx$. Here, u and v are two parameters, and are interpreted as stimulation from new external inputs. In the absence of stimulation, the potential function appears as in Figure 15.10a, with just one stable point corresponding to point P or Q in Figures 15.8c and 15.9. If u decreases from 0 to −1, while v stays fixed at 0, the potential curve is as shown in Figure 15.10b. The stable point has become an unstable point, like N in Figures 15.9 and 15.8c, and it is bifurcated into two new stable points, such as P and Q in Figure 15.8c. The symmetry of Figure 15.10a is broken when replaced by Figure 15.10b, which has different symmetries.

Preparedness of a mind for certain input structures corresponds to the presence of stable potential wells that can serve to attract them. Since inputs being attracted to and falling into the wells are assumed to deepen them, repeated exposure to similar structures makes it easier to fit them in (i.e., to comprehend the next instance of a similar structure).

The presence of parameters such as u and v guarantees the existence of niches for the stable points P and Q, but these niches will not be populated until stimuli to trigger bifurcation occur. What, then, is the genesis of new niches? A plausible position is that the exact structure of the potential function V(x) is not known, but nonetheless is at least circumscribed by the environment in which a thinker operates, so that it gradually becomes known. Another position is that as the environment changes, these functions emerge according to unpredictable complex processes. Nonetheless, a law may govern such change, at least at a global or macroscopic level. Such a law is best regarded as a metaphor modeled after thermodynamics principles. Processes analogous to cooling increase the order of local regions or points (structures) in knowledge space at the same time that processes analogous to heating increase the number of niches. Both processes increase complexity, but make the world more interesting, perhaps at a faster rate than that at which science is able to enrich its knowledge space to comprehend it all.

Conclusion

Hopefully, the preceding discussion has presented some fresh and fruitful ideas about how a mind could organize and add to its knowledge. The main idea is that the basic units of cognition are knowledge or concept structures with varying degrees of order. A structure exists simultaneously at several levels. It is what different representations of the same concept have in common. A canonical representation may be the language of topological complexes, or more simply, the language of graphs. At one level, a concept is a node in some graph. Its meaning is in the way that it is connected with other nodes. The connections or edges are themselves concepts, denoted by nodes at another level. And each node can be replaced by a graph structure one level down. Such a structure is highly ordered to the extent that it has many symmetries, and requires, at best, a short code to describe it.

At the highest level, such structures are points in a knowledge space, regarded as a continuous topological surface, with a dent or potential well at that point. If similar structures at nearby points on the surface are encountered as environmental stimuli, they are attracted and fall into that well, deepening it as well as the path to the well. The resulting deformations of the surface contain points that are stable bifurcation points and unstable points. The symmetries of the potential wells reflect the symmetries in the structures that map into them. Accumulation of information, viewed as the decrease in uncertainty that accompanies the emergence of order in local structures, destroys symmetries in global structures. This makes the world of knowledge more complex, but also more interesting.

Thinking is regarded as the control of attention to nodes in some knowledge structure. In a highly ordered structure, thinking is fast and imposes little cognitive strain. Conceptual combinations are favored. The thinker's mind is prepared to recognize many structures and to anticipate new ones. Comprehension and discovery are facilitated. A less ordered global structure comprised of more highly ordered local structures is interpreted as a prepared mind, one that can make discoveries more readily and quickly.

The ability of a thinker to generalize, classify, and abstract from its experience—which is central to cognition—is reflected in procedures that recognize similarities among structures. Again, similarity recognition results in a structure with many symmetries that reflect the similarities. Hypotheses or beliefs are modeled as composite structures, and the same procedures that form structures apply to the generation and modification of beliefs.

That the structures discussed here correspond to neuronal groups is a speculation, but one that may lead to fruitful laboratory science. Beyond that, the discussion has been metaphorical, and this may be appropriate for this topic at this time. Whether there are a few simple, powerful laws to account for the mind's surprising ability to discover and imagine is, at present, a matter of controversy. This chapter presented the position that there are such laws. This belief is partially justified by what appears a general trend in the sciences toward concern with finding unifying principles. In physics, Einstein's lifelong vision that all the known forces (now but three) can be unified is coming closer to fruition with the concept of gauge symmetries. Lowering of the temperature in the universe—the average is now 3° K—is believed to have destroyed the symmetries that a single force would display at a higher temperature. Regarding the development of the nervous system and connectivities in the brain, evidence now appears to support the view that a single, modifiable molecule, N-CAM, is responsible for intercellular bonds and all nerve interactions. Under the influence of other molecules, it shapes patterns and mediates neuronal recognition and specificity. The simple and general notion of an ordered knowledge structure (rather than a production rule) and procedures that are themselves structures for generating, modifying, and using structures may lead to such simple, powerful laws in cognitive science.

This chapter has featured some powerful new ideas, such as that of an ordered knowledge structure, and suggestions for simple laws, such as the procedures for modifying and reorganizing knowledge structures. It has argued that these can account for many cognitive processes. It has drawn upon and made contact with many more ideas in the literature than have been cited, notably work on conceptual structure and its evolution, search theory, and the Boltzmann machine. Even such connections as were cited could not be elaborated more fully without greatly lengthening the chapter. An interesting historically and philosophically oriented analysis of the literature (Flanagan 1984) concludes that the "overwhelming questions about who we are and who we are to become" are unlikely to have one simple, neat, and tidy answer. Of course, we ask much simpler questions, and look not to one discipline, but to a novel synthesis of several rapidly advancing scientific frontiers. The reader can readily see connections between this and the other chapters in this volume, which represent (in our opinion) the most promising—but not the best known—advances in cognitive science.

Appendix

What Else a Person Thinks About a Straight Line (adapted from free associations by Gloria Mark, an artist now concerned with explicating visual discoveries)

1. A direction of gravity or verticality.
2. Low tension.
3. High tension, like a taut rubber band.
4. Very strong image (like an iron bar).
5. Immutable, feeling it is stable, expressing stability.
6. Positive space (i.e., figure in a figure-ground context).
7. Low weight.
8. Floats on the page (unless it were drawn from one edge of the page to another).
9. Two-dimensional (as opposed to going backward in perspective).
10. Inorganic.
11. A drawing element that can be twisted or bent (has elasticity).
12. Shows expansion from a center point in two directions.
13. A support for something: (a) if vertical, it is supported by its end; (b) if horizontal, supported in the center.
14. A slash in the field (in this sense, it is viewed as negative space).
15. A cross-sectional view of a 3-dimensional shape (like a piece of paper).
16. An angle at 180 degrees.

17. A cross-sectional view of a semi-circle, or wavy line that is rotated to the point where it appears straight.
18. A unit of measure.
19. A symbol for: number 1, a worm, a human figure.
20. An indention in the paper.
21. A ray of light.
22. A tally mark.
23. The horizon.
24. A slash of rain.
25. A stroke in a character.
26. Can be indefinitely extended.
27. Is self-similar.
28. Set of real numbers.
29. Everywhere dense.
30. Contains a continuum of points.
31. Is on one-on-one correspondence with every other line segment.
32. Has length.

Notes

1. Pylyshin (1982) considers cognitive science to consist of the 6 subdomains: philosophy, psychology, linguistics, computer science, neuroscience, and anthropology. Of the 15 possible couplings between these six, 11 are established, such as neuropsychology and neurolinguistics. Bobrow and Collins (1975), on the other hand, used it to build upon six fields, of which philosophy, psychology, linguistics, and computer science are common to Pylyshin's analysis, but artificial intelligence and education took the place of neuroscience and anthropology. I would include artificial intelligence under computer science and adopt Pylyshin's grouping, with education added, although it would be highly desirable to maintain AI's connection with psychology, which it has influenced significantly. For this reason, a strong case can be made for AI to exist as a discipline independently of computer science and psychology, perhaps as a key discipline in the broader field of information science. It is also noteworthy that AI is sometimes viewed to have two components: knowledge engineering and cognitive science. (But that does not seem to be as appropriate.)

2. These early programs also tried to produce hypotheses about the simulated learner's self, the actions it could take, and what aspects of the environment it could control, even about its ability to form hypotheses. Therein, we thought, lay the seeds of the idea of consciousness. To generate linguistic discourse and comprehension was believed to require the formation of hypotheses about the ability of other environmental objects to form hypotheses—including hypotheses that reflect intentions. But if performance of a computer program justifiably could be so interpreted—in the sense that it could not be distinguished from behavior in persons called by the same name in a Turing test—we did not believe it possible to validly claim simulation of a human learner.

3. To some extent, the idea of the structures we discuss has precursors in the history of semantic nets (Israel and Brachman 1981). Some of the early papers by Stevens, Quillian, and Doyle had the germ of the idea; Quillian's model human verbal associations as a network search by a spreading activation pattern. At the level of physical embodiment, there is also a resemblance to perceptionlike structures begun by Rosenblatt and now being significantly extended in new directions by Edelman and his colleagues.

4. That is one reason that what we call structures here are not well-formed formulas in an applied predicate calculus.

5. Such is the situation in clinical medicine, and that is perhaps a reason why it is not a science and a basis for the belief that expert systems to aid medical practitioners require numerous heuristics rather than a few powerful ideas or laws.

Bibliography

Andreae, J.H. (ed.), 1974–1977: *Man-Machine Studies,* Electrical Engineering Reports, University of Canterbury, Christchurch, New Zealand.

Astrahan, M.M., et al., 1976: System R—relational approach to database management, *ACM Transactions Database on Database Systems,* 1, 2, June, 97–137.

Atkin, R., 1974: *Mathematical Structures in Human Affairs,* Heinemann, London.

Barr, A., and E. Feigenbaum, 1982: *The Handbook of Artificial Intelligence,* W. Kaufmann, Los Altos, Calif.

Blum, L., and M. Blum, 1975: Toward a mathematical theory of inductive inference, *Information and Control* 28:122–155.

Bobrow, D.G., and A. Collins, 1975: *Representation and Meaning,* Academic Press, New York.

Brachman, R., E. Ciccarelli, N. Greenfield, and M. Yonke, 1978: *K-LONE Reference Manual,* BBN Report No. 3848, Bolt, Beranek and Newman, Cambridge, Mass.

Bruner, J.S., J.G. Goodnow, and G.A. Austin, 1956: *A Study of Thinking,* Wiley, New York.

Chaitin, G.J., 1977: Algorithmic information theory, *IBM Journal of Research and Development* 21:350–359.

Davis, R., 1982: Expert systems—where are we? and where do we go from there?, *AI Magazine* 3:3–22.

Davis, R., and D.B. Lenat, 1982: *Knowledge-based Systems in Artificial Intelligence,* McGraw-Hill, New York.

de Mey, M., 1982: *The Cognitive Paradigm,* Reidel, Dordrecht, Holland.

Doran, J., 1968: Experiments with a pleasure-seeking automaton, *Machine Intelligence,* Edinburgh University Press, Edinburgh, Scotland.

Elithorn, A., and R. Banerji, 1982: *Artificial and Human Intelligence,* Elsevier, New York.

Flanagan, O.J., 1984: *The Science of the Mind,* Bradford, Cambridge, Mass.

Gardner, Martin, 1983: The computer as scientist, *Discover*, June.

Gold, E.M., 1971: Universal goal-seekers, *Information and Control* 18, 5: 395–403.

Groner, R., M. Groner, and Bischof (eds.), 1982: *Methods of Heuristics*, Lawrence Erlbaum, Princeton, N.J.

Hantler, S., and M. Kochen, 1973: ASP: a system using stored hypotheses to select actions, *Journal of Cybernetics* 3:1–12.

Israel, D.J., and R.J. Brachman, 1981: Distinctions and confusions: a catalogue raisonne, *International Joint Conference on Artificial Intelligence*, Vol. 1, 452–459.

Kochen, M., 1960a: *Cognitive Mechanisms*, IBM Report, IBM Research Center, Yorktown Heights, N.Y.

Kochen, M., 1960b: An experimental study of strategies in hypothesis-formation by computers, *Transactions of the 4th Symposium of Information Theory*, Butterworth, London.

Kochen, M., 1962: Adaptive mechanisms in digital concept-processing, *Proceedings of the Joint Automatic Control Conference*, American Institute of Electrical Engineers.

Kochen, M., 1971: Cognitive learning processes—an explication, in N.V. Findler and B. Meltzer (eds.), *Artificial Intelligence and Heuristic Programming*, Edinburgh University Press, Edinburgh, Scotland, 261–317.

Kochen, M., 1974: Representations and algorithms for cognitive learning, *Artificial Intelligence* 5:199–216.

Kochen, M., 1975: An algorithm for forming hypotheses about simple functions, *Proceedings of the Milwaukee Symposium on Automatic Computation and Controls*, Milwaukee, Wisc.

Kochen, M., 1982: An evolutionary approach to hypothesis and concept formation, in R. Groner, M. Groner, and Bischof (eds.), *Methods of Heuristics*, Lawrence Erlbaum, Princeton, N.J., 37–67.

Kochen, M., and J. Stark, 1978: Representation and formation of hypotheses in learning programs, *Proceedings of the International Conference on Cybernetics and Society*, IEEE, Tokyo.

Kugel, P., 1977: Induction, pure and simple, *Information and Control* 35:76–366.

Lenat, D.B., 1976: *AM—An Artificial Intelligence Approach to Discovery in Mathematics As Heuristic Search*, Memo AIM-286 Stanford Artificial Intelligence Laboratory, Ph.D. thesis, Computer Science Department, Stanford University, July.

Lenat, D.B., 1982: Heuristics: theoretical and experimental study of experimental rules, *Proceedings of the AAAI-82*, Pittsburgh, Pa., 159–163.

McGill, M.J., 1976: Knowledge and information spaces, *Journal of the American Society of Information Science*, 27, 4, July–August, 205–210.

Mulsant, B., and D. Servan-Schreiber, 1983: *Knowledge Engineering—A Daily Activity on a Hospital Ward*, Stanford University Report, May.

Pylyshyn, Z.W., 1982: Information science: its roots and relations as viewed from the perspective of cognitive science, *Knowledge*, December.

Shipman, D.W., 1981: The functional data model and the data language DAPLEX, *ACM Transactions on Database Systems* 6, 1, March, 140–173.

Shortliffe, E., and R. Duda, 1983: Expert systems research, *Science* 220:261–267.

Solomonoff, R.J., 1964: A formal theory of inductive inference, *Information and Control,* March, 1–22, and June.

Solomonoff, R.J., 1982: Private communication.

Thagard, P., 1983: *Conceptual combination—a frame-based theory,* paper presented to Society for Philosophy and Psychology, Boston.

Winston, P., and B. Horn, 1981: *LISP,* Addison-Wesley, Reading, Mass.

FIGURE 15.1 A configuration of disks on pegs

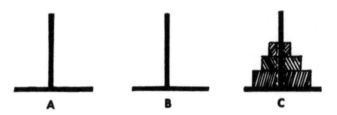

FIGURE 15.2 A configuration to be obtained as a goal

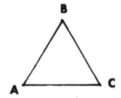

FIGURE 15.3 Configuration transition diagram for one disk

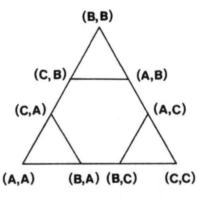

FIGURE 15.4 Configuration transition diagram for two disks
[Notation: (smaller disk, larger disk)]

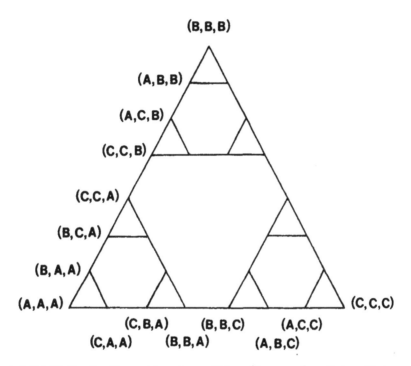

FIGURE 15.5 Configuration transition diagram for three disks
[Example: (B,C,A) means smallest disk on peg B, mid-sized disk
at C, largest disk at A.]

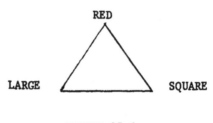

FIGURE 15.6a

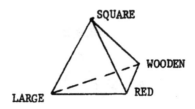

FIGURE 15.6b

FIGURE 15.6 Representation by topological simplexes
(Q-analysis)

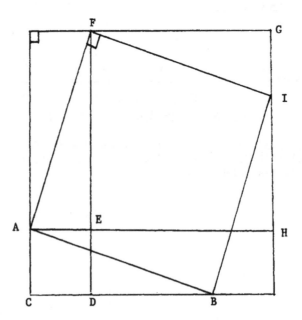

FIGURE 15.7 The Pythagorean Theorem and its proof

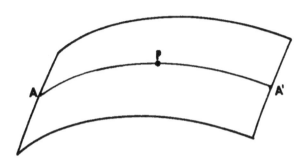

FIGURE 15.8a A possible two-dimensional knowledge space as
a surface in Euclidean 3-space

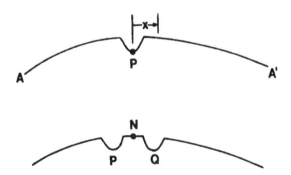

FIGURES 15.8b, 15.8c Cross sections of two-dimensional
knowledge spaces cut by a vertical plane through A and A'

FIGURE 15.8 A possible representation of a knowledge space

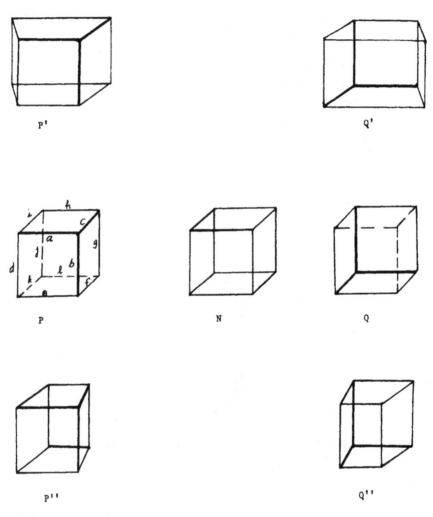

FIGURE 15.9 Examples of conceptual structures and transformations among them

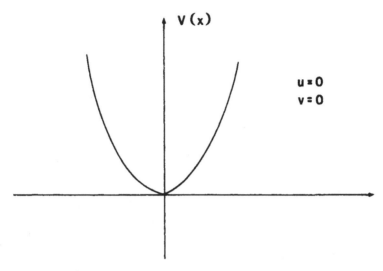

FIGURE 15.10a

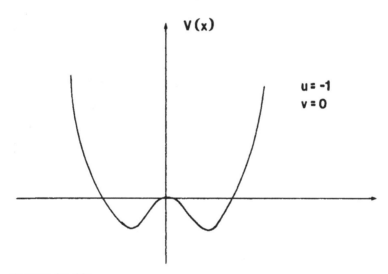

FIGURE 15.10b

FIGURE 15.10 Potential wells for niches in a knowledge space
$[V(x) = x^4/4 + ux^2/2 + vx]$

16. Conclusion

The overall theme of this volume can be summarized succinctly as "intelligent behavior." The authors presented both natural (biological) and artificial (computer) approaches to several facets of intelligent behavior. These included possible models for intelligent behavior: decision theory (Cooper in Chapter 2), biological intelligent behavior as contrasted with the usual concept of computing (Yates in Chapter 3), and Darwinian type selection in learning systems (Reeke and Edelman in Chapter 4).

Several authors then focused on networks of neurons (Hastings and Waner in Chapters 5 and 7) and on processors (Resnikoff in Chapter 6). In fact, neural networks can be efficiently modelled on massively parallel computers like the Connection Machine or the National Aeronautics and Space Administration's Massively Parallel Processor (MPP) by mapping one neuron to each processor. The concepts of selection-based and evolutionary learning were further developed in Greenspan's work on the development of the nervous system (Chapter 8) and Freeman's chapter (9) on synergetics in the brain, which demonstrated the information-processing power of multilevel and cooperative behavior.

Conrad, Kampfner, and Kirby (Chapters 10, 11, and 12) considered evolutionary learning in more detailed formal neural models. In particular, the complex behavior of individual neurons adds additional processing power to the network-level behavior.

The last three chapters dealt with the goal of information processing, intelligence, and/or cognition: problem solving. In Chapter 13, Solomonoff discussed the complexity of real-life problems in cognition from the basis of computational complexity theory. Lindsay and Kochen, in Chapter 14, considered the problems of backtracking, and found a basis for the development of novel ideals within a system, characteristic of intelligent behavior. Kochen concluded by considering knowledge structures from the viewpoint of cognitive science, gaining informative insights for future intelligent systems (Chapter 15).

The introduction (Chapter 1) considered these results from the primary viewpoint of the convergence of computer science and biology in the theory of cognitive science. These concluding remarks will show how these same results extend the development of computer-based information processing, and represent powerful new approaches to some classic problems.

It is appropriate to begin by reviewing some facets of the development of the theory of computing and connections with intelligence, as well as hardware and software.

By the 1940s, the major foundations of theoretical computer science, early digital computers, and the basic model for formal neurons were all in place. The concept of algorithm as a well-defined procedure for computation was central. According to the Church-Turing thesis, any computable function could be computed by a suitable algorithm on an abstract formal computer called a Turing machine. A Turing machine consists of a finite automaton as processor and a read-write tape as additional memory or storage. The processor produces an output symbol from its current state and input symbol, and also controls data loading and storage. Both von Neumann's first computers and current standard digital computers are implementations of the Turing framework.

At about this time, McCulloch and Pitts (1943) introduced a class of formal neurons consisting of threshold logic circuits. It is easy to show that networks of McCulloch-Pitts neurons can simulate universal Turing machines, and thus that biological organisms can compute. However, McCulloch-Pitts neurons are a significant oversimplification of neural dynamics as presently understood, and thus do not answer the question: Can computers think?

We recall Godel's (1931, cf. 1952) incompleteness theorem which demonstrated important limits on decision making in formal systems. Roughly, Godel's theorem states that any consistent, sufficiently large system must be incomplete in the sense that the system contains formally undecidable statements. A statement in a system is formally undecidable if one cannot determine the truth or falsehood of that statement within that system. More precisely, if statement S is formally undecidable, then one can adjoin either of the two statements "S is true" or "S is false" to the system and still obtain a consistent system. The key to the proof of Godel's theorem is the clever use of self-reference: the use of the system to make statements about itself. Self-reference also appears in the simpler Russell's paradox, which may be stated as follows: "Let S be the set of all sets which do not contain themselves as elements. Is S an element of itself?" Both yes and no answers cause paradoxes.

Both Russell's paradox and Godel's theorem appear to limit the power of Turing machines. The limitations implied by self-reference come to clearest focus in the Turing machine halting problem. Question: Given a Turing machine T and input I (I represents the initial state of the read-write tape of T), will T halt on input I? One can show that this question cannot be answered by a Turing machine through the clever use of the existence of universal Turing machines, Godel numbering, and self-reference. Despite these limitations, von Neumann (in notes later written by Burks, 1970) demonstrated the theoretical possibility of self-reproducing Turing machines through the concept of cellular automata. A cellular automaton is an array of automata, each of which receives inputs from its neighbors. The Connection Machine and the MPP are physical realizations of cellular automata. It is worth noting that von Neumann's self-reproducing cellular automata mimic many aspects of biological organisms; namely, distributed dynamics and the existence of a separate genotype, which describes the construction of the phenotype.

Although the Turing framework has proven correct for formalizing algorithmic computations, some of our authors—in particular, Yates—question whether biological systems follow the Turing model. Hofstader (1979), as well as several authors in this volume (especially Freeman and Greenspan), describe an escape route from the Russell-Godel-Turing limit via multilevel computing.

Von Neumann (1956) also considered another extension to the Turing framework. This was a stochastic Turing machine, whose computations involve random factors, as well as the desired algorithm. Stochastic automata are one starting point for the work of Hastings and Waner on evolutionary learning in this volume.

Although (the concepts of) both cellular and stochastic automata have now made profound impacts, hardware limitations prevented a more immediate effect.

In the late 1940s, time vacuum tubes had just replaced relays as the logic elements of digital computers. Besides being relatively slow, vacuum tubes were also subject to random (Poisson) failures, with a typical mean lifetime of perhaps 8,000 hours. This means that in a computer with 16,000 vacuum tubes (roughly the size of the Eniac), the mean time between failures was about 30 minutes. Since computations might require much more time, these failures clearly were a source of errors to be avoided. This represented the important beginning of fault-tolerant design for computers. However, it seems that the presence of such difficulties dictated the path of the theory of stochastic automata toward a position that "stochastic behavior represents a class of errors to be avoided." This viewpoint is also consistent with the developing theory of the algorithm and with Shannon's work on the problem of communication over a noisy channel.

An exponentially rapid increase in computer speed occurred from the 1940s to the 1980s as vacuum tube logic elements were successively replaced by transistors, small integrated circuits, and very large-scale integrated circuits. These developments also had a profound impact on the development of machine intelligence.

There are many problems whose solution appears to demonstrate intelligence. Consider, for example, the problem of finding the next move in a chess game. One might consider the decision tree formed of all possible legal moves and sequences of responses, continuing up to some desired depth of search. This tree is very large, and the number of vertices grows exponentially with the depth of the search (a familiar situation in the theory of computational complexity, cf. Aho, Hopcroft, and Ullman 1974; Garey and Johnson 1979). Human expert chess players use their intuition and experience to select a limited number of possible moves and consequent responses resulting from such moves, thus limiting the size of the tree, and then perform detailed analysis on this relatively limited tree. Because of the difficulty of describing intuition algorithmically, traditional computer approaches to chess have relied instead on very fast processors to handle a much larger decision tree.

The approach of evolutionary computing and cognitive science is to try to develop systems which "learn" the required intuition in a suitable sense. It is hoped that these approaches will combine the power of high speed computers with that of biological systems. We remark that evolutionary computing requires suitably random searches, which can be efficiently modelled by stochastic automata. In addition, massively parallel computers are physical realizations of cellular automata. Thus, the two facets of von Neumann's vision extending classical automata are realized in many present approaches.

We conclude this historical overview with three recent developments. The first involves a procedure called simulated annealing introduced by Metropolis et al. (1953) in order to solve chemical equations of state. Since that time, simulated annealing has been used to rapidly find approximate solutions to a wide variety of difficult (NP-complete, cf. Aho, Hopcroft, and Ullman 1974; Garey and Johnson 1979) optimization problems (see the survey article of Kirkpatrick, Gelatt, and Vecchi 1983). The problem of evolutionary searching in a protein space (Maynard-Smith 1970) and many similar natural problems are at least that difficult. Simulated anealing has been a fundamental technique in much recent work on neural networks, including Hopfield memory (1982), the Boltzmann machines of Hinton, Sejnowski, and Ackley (1984), and the evolutionary learning machines of Hastings and Waner (this volume). In addition, Conrad's (1979) theory of evolutionary search and Conrad's evolutionary computing (Chapters 10–12, this volume) implicitly use annealing algorithms.

Simulated annealing is based on physical annealing, in which heating and cooling of a metal is used to reduce the potential energy of internal stresses. The dynamics of annealing interpolate between those of random search (at very high temperatures) and those of gradient systems (at temperature zero). In particular, the low-temperature dynamics are close to those of gradient systems, which rapidly approach local minima. However, the small amount of randomness still present allows the state to avoid the trap of local, nonglobal minima. High-temperature dynamics rapidly sample even a very large state space. It is this combination of dynamics, together with suitable control over the temperature, which makes annealing work well.

In Hastings and Waner (1984), the authors argued that biological systems may directly effect annealing through their internal dynamics. This allows vast increases in potential computing power.

The second major recent development is that of very powerful expert systems developed by Feigenbaum and many others. The systems use large knowledge bases and automated reasoning (in most cases, using von Neumann machines, although there are recent powerful extensions to the Connection Machine) to duplicate the performance of expert-level information processing in fields as diverse as medicine, prospecting, and chess.

The third major event is the development of large parallel computers. Not only have such machines allowed significant speed increases over conventional computers, but they have focused on several related issues. First, communication bandwidth may be the ultimate limit to computational speed. This requires distributed computing, as well as the existence of a (possibly large) number of suitable computing primitives. Second, optimal utilization of these resources may require asynchronous computing, with the consequent loss of structural programmability. In fact, we conjecture that biological systems utilize side effects in an essential way, in sharp contrast to structurally programmable digital computers in which the role of side effects is limited. If, as seems likely from the material in this volume, biological systems are efficient users of computational resources, then one may see a radical shift in some directions of artifical intelligence from powerful data-based systems toward evolutionary learning systems.

The chapters and major themes of this volume represent logical and potentially powerful extensions of these themes. Evolutionary learning is a natural application or realization of annealing. This is true, at least in part, of the essential similarity between stochastic decision theory and mutation and selection in Darwinian evolution (Chapter 2), selection as a mode of learning in the nervous system (Chapter 4), and the development of the nervous system (Chapter 8). Conrad, Hastings, Kampfner, and Waner directly addressed evolutionary learning in this manner. It appears reasonable that the immense power of annealing in generating useful solutions to complex optimization problems will have parallels in the use of evolutionary learning to attack similarly complex learning problems such as those described in Chapter 13.

The concept of knowledge-based expert systems is extended in Greenspan's theory of biological information storage and realization and Kochen's study of order and disorder in knowledge structures. Lindsay and Kochen discussed the appearance of novelty, which may correct the lack of what may be termed creativity in expert systems. Many authors covered the general management of information.

Parallelism was explored in work on neural networks, as well as concurrent processors (Chapter 6). Conrad and Kampfner considered the computational contributions of the rich structure of neurons interacting in these networks. It is at least plausible that evolutionary algorithms may make better use of nonconcurrent parallel processor than traditional algorithms.

In conclusion, although biological systems may not compute in the sense of Turing (see Chapter 3), they appear to provide appropriate models for the theory and practice of cognition, as well as its machine implementation.

Bibliography

Aho, A.V., J.E. Hopcroft, and J.D. Ullman, 1974: *The Design and Analysis of Computer Algorithms*, Addison-Wesley, Reading, Mass.

Burks, J.W., 1970: *Essays on Cellular Automata*, University of Illinois Press, Urbana, Ill.

Garey, M.R., and D.S. Johnson, 1979: *Computers and Interactibility: A Guide to the Theory of NP-Completeness*, W.H. Freeman, San Francisco.

Godel, K., 1931: Uber formal unentscheidbare Satze der Principia Mathematica und verwandter System I, *Monatschafte fur Mathematics und Physics* 38:173–189.

Godel, K., 1962: *On Formerly Undecideable Propositions*, Basic Books, New York.

Hastings, H.M., and S. Waner, 1984: Low dissipation computing in biological systems, *BioSystems* 17:241–244.

Hinton, G.F., T.J. Sejnowski, and D.H. Ackley, 1984: *Boltzmann machines—constraint satisfaction networks that learn* (preprint), Carnegie-Mellon University, Pittsburgh, Pa.

Hofstad, D.R., 1979: *Godel, Escher, Bach: An External Golden Braid,* Vintage Books, New York.

Hopfield, J.J., 1982: Neural networks and physical systems with emergent collective properties, *Proceedings of the National Academy of Sciences USA* 79:2554–2558.

Kirkpatrick, S., C.D. Gelatt, Jr., and M.P. Vecchi, 1983: Optimization by simulated annealing, *Science* 220:671–680.

Maynard-Smith, J., 1970: Natural selection and the concept of a protein space, *Nature* 255:563–564.

McCulloch, W.W., and W. Pitts, 1943: A logical calculus of the ideas imminent in nervous activity, *Bulletin of Mathematical Biophysics* 5:115–133.

Metropolis, N., A. Rosenbluth, M. Rosenbluth, A. Teller, and E. Teller, 1953: Equations of state calculations by fast computing machines, *Journal of Chemical Physics* 21:1087–1091.

von Neumann, J., 1956: Probabilistic logic and the synthesis of reliable organisms from unreliable components, in C.E. Shannon and J. McCarthy (eds.), *Automata Studies,* Princeton University Press, Princeton, N.J., 43–98.

About the Contributors

Manfred Kochen is Professor of Information Science and Research Mathematician at the University of Michigan's Mental Health Research Institute. He is also Adjunct Professor of Computer and Information Systems at the university's Graduate School of Business Administration. Since 1984, he has chaired the Sociotechnological Systems Area in the Urban, Technological, and Environmental Planning Ph.D. Program. He has expertise in problem formulation and solving, system analysis design, creative modelling, and all facets of cognitive processing and information management. He earned a B.S. in General Science at the Massachusetts Institute of Technology, and he received his M.A. in Mathematics and Ph.D. in Applied Mathematics from Columbia University. Honors include the Award of Merit from the American Society for Information Science and a postdoctoral Ford Fellowship, and he has written and edited numerous articles and books.

Harold M. Hastings is Chairman of the Department of Mathematics, Professor of Mathematics, and Adjunct Professor of Computer Science at Hofstra University, and he is a member of the NASA Massively Parallel Processer national working group. His particular areas of concentration are neural nets, artificial intelligence, math models, and topology. He has received notable academic support for his work, including two National Science Foundation grants and a Woodrow Wilson National Fellowship Foundation grant, as well as a number of awards and honors, including the Hofstra Distinguished Service Award. He has organized several conferences on artificial intelligence, evolutionary computing, and homotopy theory. Hastings has lectured and published extensively on evolutionary learning, fractal models, shape theory, and systems stability. He received a B.S. summa cum laude from Yale University, and his M.A. and Ph.D. degrees from Princeton University.

Michael Conrad is Professor of Computer Science and Biological Sciences at Wayne State University. His special interests are biological information processing and brain theory, and computational modelling of complex biological systems. He has written over 100 articles on the topics of biological information processing and molecular computing, evolutionary theory and ecology, and the comparative aspect of information processing in biological systems and artificial computers. He has also written a book, *Adaptability: The Significance of Variability from Molecule to Ecosystem* (Plenum Press, New York, 1983). He has received several fellowships and honors and has served as a visiting scientist for various institutions in the USSR, Japan, and India. Conrad earned his A.B. from Harvard University and received a Ph.D. from Stanford University.

William S. Cooper is Professor of Library and Information Studies at the University of California at Berkeley. He has held various faculty and research positions at prominent institutions, including Acting Director of the Institute for Library Research, a state-wide research unit of the University of California, and Miller Professor at the Miller Institute for Basic Research in Science in Berkeley. Cooper earned a B.A. in Mathematics from Principia College, he received a M.Sc. in Mathematics from the Massachusetts Institute of Technology, and he received a Ph.D. in Logic and the Methodology of Science from the University of California at Berkeley. He has been awarded several fellowships and other academic honors; twice his work was named the Best Paper of the Year by the American Society for Information Science. He has published extensively on the subjects of information retrieval, decision theory, logic linguistics, and utility and probability theory.

Gerald M. Edelman is an eminent medical researcher whose career has been distinguished by many advanced degrees in science and medicine, membership in many scientific and honorary societies, and a number of awards and honors, including the Nobel Prize for Physiology or Medicine in 1972. He is Director of the Neurosciences Institute and Scientific Chairman of the Neurosciences Research Program. Among the areas Edelman is currently studying are embryonic cell interactions (particularly those involving cell adhesion molecules), developmental genetics, higher brain functions, and the construction of recognition automata. His major theoretical interests are concerned with a new theory of brain function known as Neural Darwinism, and his experimental work revolves largely around efforts to explore its implications.

Walter J. Freeman is Professor of Physiology at the University of California at Berkeley. His area of specialization is neurophysiology, and he has written numerous articles on the mathematical biophysics of the cerebral cortex, its structure, and its nonlinear dynamics subserving behavior. He has also written a book, *Mass Action in the Nervous System* (Academic Press, New York, 1975). Freeman received his M.D. degree from Yale University, and he was named a Guggenheim Fellow in 1965 and Titulaire de la Chaire Solvay, Université Libre de Bruxelles in 1974. He is a member of a number of research and professional organizations.

Ralph J. Greenspan is Associate Member of the Roche Institute of Molecular Biology, Department of Neurosciences. Previously, he was Assistant Professor of Biology at Princeton University. His area of expertise is genetic neurobiology as applied to genetic information for development of the nervous system in mice and in the fruit fly. He has previously been instructor in neurobiology at both Cold Spring Harbor Laboratory and the Catholic University of Chile. He has been awarded several fellowships and honors, and he has published a number of papers on neurogenetics and neurobiology. Greenspan received his B.A. and Ph.D. in Biology from Brandeis University, and he was a Postdoctoral Fellow in Neurobiology at the University of California at San Francisco.

Roberto R. Kampfner is Assistant Professor of Computer Science at Wayne State University. His work lies in the areas of biological information processing, evolutionary programming, and information systems and organizational information processing. He has written articles dealing with computational modelling of evolutionary learning processes, properties of enzymatic neuron networks, and information system requirements. Kampfner earned a M.S. in Systems Analysis from the London School of Economics and Political Science, and he received a Ph.D. in Computer and Communication Sciences from the University of Michigan.

Kevin G. Kirby is a graduate student in the Department of Computer Science at Wayne State University, working under Dr. Michael Conrad. His field of specialization is reaction-diffusion models for neural membranes and intraneuronal activity.

Robert K. Lindsay is Research Scientist at the University of Michigan School of Medicine. His field of specialization is computer science and artificial intelligence. He earned a B.S. from Carnegie-Mellon University, a M.A. from Columbia University, and he received his Ph.D. from Carnegie-Mellon.

George N. Reeke, Jr., is Associate Professor at The Rockefeller University in New York City. His special research interests are biological pattern recognition and classification, neural network modelling, protein structure and function, protein crystallography, and crystallographic computing. Reeke earned a B.S. in Chemistry at the California Institute of Technology, and he received his M.A. in Chemistry and Ph.D. in Physical Chemistry from Harvard University. He has received various awards and honors, including several fellowships. Among his many publications are papers on the topics of selective networks, associative memory, cystallographic computing, and recognition automata.

Howard L. Resnikoff is President of Aware, Inc. Previously, he was Vice-President and Director of Research at Thinking Machines Corporation, of which he is a founder. He has recently completed a term as Visiting Scientist at the Massachusetts Institute of Technology's Center for Biological Information Processing. Previously, he was an Associate Vice- President at Harvard University, where he taught in the Departments of Psychology, Applied Mathematics, and Computer Science. Resnikoff was the first Director of the Division of Information Science and Technology at the National Science Foundation, and he has also served as Chairman of the Department of Mathematics at the University of California at Irvine. He has been a Member of the Institute for Advanced Study and has received the Alexander von Humboldt Foundation U.S. Senior Scientist Award.

Ray J. Solomonoff is Principal Scientist at Oxbridge Research. His expertise lies in the areas of artificial intelligence, inductive inference, and learning. He is best known for his discovery of algorithmic probability and has written articles treating the application of algorithmic complexity to induction, and comparative complexity-based induction systems. He is a member of AAAS, the Institute of Electrical and Electronics Engineers, and the Association for Computing Machinery. He received his M.S. in Physics from the University of Chicago, and he was a staff member of the Artificial Intelligence Lab at the Massachusetts Institute of Technology.

Stefan Waner is Associate Professor of Mathematics at Hofstra University. His research focuses on the topics of equivariant algebraic topology and artificial intelligence via evolutionary learning networks. He is also involved in directing student work in computing science based on neural network model implementation and in designing mathematics courses for computer graphics and business students. He has published a number of papers dealing with evolutionary computing and algebraic topology. Waner completed B.Sc. degrees in Mathematics at the University of the Witwatersrand, Johannesburg, South Africa, and the University of Liverpool, England. He received his M.S. and Ph.D. in Mathematics from the University of Chicago, and he was awarded several prizes and honors for outstanding scholarship.

F. Eugene Yates is Director of the Crump Institute for Medical Engineering, Professor of Medicine, Crump Professor of Medical Engineering, and Professor of Chemical Engineering at the University of California at Los Angeles. He is also Consulting Principal Scientist for the ALZA Corporation. His work focuses on issues in medical engineering, physiological monitoring, and dynamic systems analysis.

His numerous articles include papers on endocrinology, metabolic physiology, systems biology, and aging. He has been honored with various awards and lectureships, he is active in many professional societies and on boards and committees, and he has organized major conferences in his field. Yates received his A.B. and M.D. degrees from Stanford University School of Medicine.

T - #0035 - 071024 - C0 - 229/152/16 [18] - CB - 9780367014254 - Gloss Lamination